V&R

Hans Schwarz

Vying for Truth – Theology and the Natural Sciences

From the 17th Century to the Present

Vandenhoeck & Ruprecht

Cover: © Edward Samuel – Fotolia.com

Bibliographic information published by the Deutsche Nationalbibliothek
The Deutsche Nationalbibliothek lists this publication in the Deutsche
Nationalbibliografie; detailed bibliographic data available online:
http://dnb.d-nb.de.

ISBN 978-3-525-54028-2
ISBN 978-3-647-54028-3 (E-Book)

Typesetting by: Konrad Triltsch, Print und digitale Medien GmbH, Ochsenfurt
Printed and bound in Germany by ⊕ Hubert & Co, Göttingen

Printed on non-aging paper

Contents

Preface 7

Introduction 10

1 The Beginning of a New Age 13

2 Theology in Retreat (19th Century) 23
  2.1 The Materialistic Attack 24
  2.2 The Evolutionistic Attack 33
  2.3 The Reaction of Theology 42

3 British Empiricism and Its Consequences (17th–19th
  century) 48
  3.1 The Battle over Innate Ideas 49
  3.2 The Design Argument 51
  3.3 The Battle over Darwin 56

4 North America's Problem with Darwin 60
  4.1 Euphoria in a Progressive World 61
  4.2 From Fear to Embrace: Protestant Theology and
    Evolution 71

5 The Continental Fortress Mentality and the Gradual
  Turnaround (First Half of the 20th Century) 87
  5.1 A Strict Demarcation (Karl Barth) 87
  5.2 The Scottish Peculiarity (Thomas F. Torrance) ... 90
  5.3 A German Outsider (Karl Heim) 94

5.4 A Roman Catholic Voice (Pierre Teilhard de
    Chardin) 97
5.5 A New Beginning of the Dialogue by the Natural
    Sciences 100

6 A Vivid Dialogue with Many Voices 108
6.1 The Grand Senior of the Dialogue: Ian Barbour .. 108
6.2 The Institutionalized Dialogue 113

7 Partners from the Natural Sciences 132
7.1 Discerning the Mind of God (Stoeger, Davies,
    Hawking, Tippler) 133
7.2 Between Rejection and Proof (Wuketits, Dawkins,
    Wilson, Kutschera, Gitt, Dembski, Scherer) 148
7.3 Human Accountability and Religious Naturalism
    (Hans-Peter Dürr and Ursula Goodenough) 167

8 Partners from Theology 172
8.1 Different Traditions (Russell, Hefner,
    Polkinghorne, McGrath, Deane-Drummond, Ijjas,
    Drees) 172
8.2 Dialogue as an Avocation (Moltmann,
    Pannenberg) 198

9 Important Issues 207
9.1 Nature or Creation? 207
9.2 Brain and Spirit 215
9.3 Responsible Shaping of the World 222

Index of Names 229

Index of Subjects 234

Preface

The philosopher and natural scientist CARL FRIEDRICH VON WEIZSÄCKER (1912 – 2007) entitled his Gifford Lectures (1959 – 1961) *The Relevance of Science*. In the first sentence of this lecture series he emphasized: "Our age is an age of science."[1] This estimate is still true today. Then von Weizsäcker put forth two theses: "1. Faith in science plays the role of the dominating religion of our time. 2. The relevance of science for our time can, in this moment of history, only be evaluated in terms that express an ambiguity."[2]

The American systematic theologian LANGDON GILKEY (1919 – 2004) described the religious dimension of science by reminding us of the contradictory image of humanity "as helpless patients in the backless hospital shift and yet as mighty doctor in the sacral white coat" from whom the patients expect "redemption" from their disease.[3] Indeed we expect from science in its applied form as technology the solution to all problems whether with regard to medicine or the environment. It is telling that the so-called Green Party in Germany condemns certain forms of applied science while it advocates other forms, for instance renewable forms of energy instead of atomic energy. The philosopher KARL JASPERS (1883 – 1969) warned a long time ago not to indulge in a so-called superstition of science. By that he

---

[1] Carl Friedrich von Weizsäcker, *The Relevance of Science: Creation and Cosmogony* (New York: Harper & Row, 1964), 1.

[2] von Weizsäcker, *The Relevance of Science*, 3.

[3] Langdon Gilkey, *Religion and the Scientific Future: Reflections on Myth, Science, and Theology* (New York: Harper & Row, 1970), 85.

meant an unlimited trust in science and that we even elevate it to the rank of a religion. Yet do the natural sciences deserve this high esteem?

In the 19th century the sciences established a solidly built and seemingly unassailable edifice of knowledge. Many people are still convinced today that the sciences, by which one understands primarily the natural sciences, give us undeniable facts which we can trust, whereas the Christian faith contains only convictions which largely cannot stand up to careful examination. Since our economic industrial progress has become more and more halting and its success is largely owed to the applied sciences, more and more people doubt at the same time whether one can blindly trust the natural sciences. In this situation it makes sense to evaluate the relationship between theology, which is the reflective faith, and the natural sciences to detect how far our trust in these is justified and how they actually relate to each other. In so doing we will first depict in bold strokes the history of the dialogue between the two and then we will turn to the most important dialogue partners today. In conclusion we will sketch out some of the present-day areas where important problems seem to arise.

Delineating the dialogue we will present it with primary focus on names and (geographical) areas. While ideas are important they are always associated with certain names, e.g., evolution and Darwin, or dialogue and Barbour. Therefore emphasis will be given on what these persons contributed to the conversation.[4] In approaching the issue of the interface between theology and science, we will naturally focus on the geographical areas in which this interface started and where it still has its center of gravity, Europe and North America. Of course, there are now also other dialogue partners in the Islamic and Buddhist world, and many other regions. Yet space did not permit to dwell on these

---

[4] Willem B. Drees, *Religion and Science in Context. A Guide to the Debates* (London: Routledge, 2010), 3, rightly observes: "Reflections on 'religion and science' take place in a cultural, social context." While these contexts dare not be neglected since they certainly shape the debates, dialogue is carried out by individual persons.

voices, important as they are.[5] Another point needs mentioning. Until very recently there has been a paucity of women represented in the dialogue. One of the first to make a real change was Antje Jackelén about whom we hear more later. This means with few exceptions the participation of women in this dialogue has not started until the 21st century and therefore women are only gradually occupying influential positions in this conversation.

The focus on the interface between theology and science has also allowed me to draw on portions of some of my earlier publications (*Creation* [Eerdmans, 2002] and *Theology in a Global Context: The Last two Hundred Years* [Eerdmans, 2005]), and to use some material from there and update and expound on it for this specific topic. I want to thank my doctoral student Mona Lisa Siacor for typing the manuscript and Dr. Terry Dohm for improving on my style. Of course, I take full responsibility for any and all omissions or infelicities in style or content.

Hans Schwarz

---

[5] For different faith traditions cf. "Part I. Religion and Science across the World's Traditions" in *The Oxford Handbook of Religion and Science*, ed. Philip Clayton, assoc. ed. Zachary Simpson (Oxford: University Press, 2006), 5-135. Another important source for the dialogue in other faith traditions is the journal *Zygon* which devotes considerable space to this dialogue.

Introduction

In October 1975 twenty-seven Nobel laureates and six theologians met at Gustavus Adolphus College in St. Peter, Minnesota to discuss before an audience of approximately 4,000 people "the future of science." Because of its Swedish roots this college has a special relationship to the Swedish committee of the Nobel Prize and stages an annual Nobel Conference for which at least one Nobel laureate is invited as the main speaker. This meeting in 1975 was especially remarkable because of the number of Nobel laureates present and also because of the topic, *The Future of Science.* Already then there was a shortage of money for research. Since in the English language science is usually understood as natural science, it was clear for the three natural scientists who presented papers that they had a special expertise in allocating the financial means and also in the use of these funds which were largely advanced by the government. Therefore they demanded that they should have a free hand. LANGDON GILKEY, a theology professor from the University of Chicago Divinity School and the only theologian who presented a paper, interjected however, that the natural sciences are threatened by a similar fate as that which has already been confronted by theology.[1]

Once, he said, theology was the queen of the sciences. Though it considers an essential aspect of human life it was pushed off its pedestal. The reason for this was its apodictic behavior. It claimed

---

[1] For the following cf. Langdon Gilkey, "The Future of Science," in Timothy C. L. Robinson, ed., *The Future of Science: 1975 Nobel Conference* (New York: John Wiley, 1977), 113.

that it must rule over all other aspects of human life and that it is the sole source of knowledge and salvation. In a similar way the natural sciences today face this confrontation, because they ascended to the throne, which prior was occupied by theology. The natural sciences, too, are indispensable and contribute essentially to life. Yet they also surround themselves with an aura of absoluteness and infallibility. Then Gilkey asked whether their claim is true, that their methods alone allow access to reality, that it is only through their research one can grasp an object completely, suggesting that the applied sciences can offer humanity affluence and security.

Most of the Nobel laureates did not like these questions posed by the theologian Gilkey. Yet he did not want to challenge the natural sciences as such because though they now threaten our survival he emphasized they are "utterly necessary for that survival."[2] We see this turnaround today, with the discussion about atomic energy. In the 1960's nuclear energy was considered the miraculous source of unlimited and cheap energy. But today more and more people have doubts about this. This change in mentality was not caused by the natural sciences but by quite different circles in the general population. This means when the future of humanity is at stake, neither the voice of the natural sciences nor that of theology has the rank of absolute validity. Both must learn that they are societal forces which are concerned with shaping and securing the future of life. Yet both of them are indispensable to the human enterprise. The natural sciences deal with the concrete shaping of the world and theology accompanies this venture in a reflective manner and connects it with the origin, meaning, and goal of this world.

In the applied sciences ethics is not an ingredient in shaping of future of the world. Similarly, faith does not figure in researching the world. Conversely, in theology the actual world receives often too little consideration while faith and conduct (ethics) are given absolute priority. Yet our societal problems, whether in technology or in the health services, can neither be solved by scientific intervention nor by ethics and faith alone. Therefore the

---

[2] Gilkey, "The Future of Science," 119.

dialogue between theology and the natural sciences gains more and more urgency. Moreover, the discoveries in physics, astrophysics, biology, brain research, and human cognition demand that the relationship between the natural sciences and faith be reconsidered. They imply questions of value and possibly also of metaphysics. On the subsequent pages we will trace this dialogue as it has evolved over the last two centuries and also point to some present-day areas where this dialogue is most urgent.

1. The Beginning of a New Age

Three figures stand out exemplifying the beginning of a new age: JOHANNES KEPLER (1571 – 1630) who discovered the laws that bear his name to calculate the planetary orbits around the sun, GALILEO GALILEI (1564 – 1642) who came into conflict with the Inquisition of the Roman Catholic Church because of his insistence on the Copernican and heliocentric worldview, and RENÉ DESCARTES (1595 – 1650) who introduced for the first time radical doubt into philosophy.

This start of a new age in which the earth lost its central place in the world and the sun emerged as the new center of the universe is often called the 'Copernican Turn'. Yet NICOLAUS COPERNICUS (1473 – 1543), an Augustinian canon from Frombork (German: Frauenburg) then part of Prussia, now of Poland, was not interested in a "Copernican Turn" but in the development of a system which would advance the classical concept of harmony to new splendor. As the American theologian HAROLD NEBELSICK (1925 – 1989) emphasized: Copernicus "clearly had no intention of abstracting his geometry from the actual motions of the heavens as such."[1] He was still too much oriented toward the Greek notion of a harmony of spheres so that he even sacrificed the accuracy of his observations for the desired elegancy of his calculations. The church, too, represented the leading opinion at that time and admonished Galilei to teach the heliocentric theory

---

[1] Harold P. Nebelsick, *Circles of God: Theology and Science from the Greeks to Copernicus* (Edinburgh: Scottish Academic Press, 1985), 237.

as a hypothesis only and not as fact.[2] It was just Johannes Kepler, who allowed his sense of harmony to be reformed by observation. He followed his mathematics and revolutionized astronomy.

According to the traditional understanding which Copernicus still represented, Kepler's planetary orbits, since they had two foci rather than one, were considered "defective" and even monstrosities. Only after a long struggle was it understood that the heavens praise the glory of God because of their creatureliness, and not because of their godlike perfection (cf. Psalm 19:1). This meant that creation does not consist of divine material but of an earthly reality which has a contingent and rational order of its own. By abandoning the idea of a Greek world harmony and emphasizing the Jewish Christian concept of the created, one could understand the material world as creation. If divine qualities could not be traced in nature then this could lead to the notion that the material world has nothing to do with God. Already in the 13th century Aristotelian philosophy had paved the way for this bifurcation or even separation of God and the world. The Islamic philosopher IBN RUSHD (Latin: Averroes; 1126–1198) wrote commentaries to virtually every work of Aristotle and exerted in the Middle Ages considerable influence on Christian scholasticism. For him it was important that through philosophy and logic a harmony was maintained between the Koran and revelation. To that effect he developed the doctrine of the twofold truth, one for the philosopher in philosophy and the other one for the masses in religion. The images and parables of the revelation of Allah in the Koran show one way to find the truth while another one is opened through timeless philosophical speculation. Though both truths seem to contradict each other the philosophic truth agrees with the religious one if correctly applied.[3] With this procedure human

---

[2] Cf. Nebelsick, *Circles of God,* 243.

[3] Cf. Karl Heim, "Zur Geschichte des Satzes von der doppelten Wahrheit," in Heim, *Glaube und Leben: Gesammelte Aufsätze und Vorträge* (Berlin: Furche Verlag, 1926), 82 ff.; cf. also R. Arnaldez, "Ibn Rushd," in *Encyclopedia of Islam. New Edition,* 3:911, in his interpretation of Averroes' "exposition of the convergence which exists between the religious law and philosophy."

reason was granted a certain freedom. This is also evident at the Reformation in the 16th century.

In the time of the Reformation the main concern was with the justifying word of God which met the individual human person in an existential way. In the subsequent century Orthodoxy endeavored to pronounce the old faith in a comprehensive and insightful way. This was not the proper time to ponder the relationship between knowledge by faith and knowledge by reason. Moreover, MARTIN LUTHER (1483–1546) had emphasized the ambivalence of reason and therefore reason ultimately could not really be supportive of faith. As the Lutheran theologian WERNER ELERT (1885–1954) noted: "The church has no interest in the different worldviews because it derives its mission from the Gospel and knows that its mission is exhausted in the proclamation of the Gospel."[4] Elert then elaborates that *"It is a manifest lie of history that Luther's authority would have hindered the advancement of the new worldview."* Unencumbered by any possible theological interests at the time of the Reformation natural scientists who advocated the Copernican worldview taught at Wittenberg University. Even GIORDANO BRUNO (1548–1600) was welcomed to Wittenberg and taught there from 1586–1588.

The natural sciences, too, had no interest to seek out theology as a dialogue partner because they were occupied with describing the contingent rational order of nature which they discovered in ever more exact details. If natural scientists, however, discussed their results in public, theology was often surprised about those results and attempted to combat these "wild theories" especially when the argument was derived from scientific insights and not just from the Bible. For this reason Galilei, Bruno, and Kepler had problems with their churches, which were intensified in the case of Galilei and Kepler since for them the natural sciences had ultimate authority in scientific matters. Yet with Bruno there were also theological problems because he wanted to limit the sole

---

[4] Werner Elert, *Morphologie des Luthertums.* Vol. 1: *Theologie und Weltanschauung des Luthertums hauptsächlich im 16. und 17. Jahrhundert* (Munich: C. H. Beck, 1965 [1931]), 371 f., for this and the following quotation.

activity of God. Even Kepler could not agree to the recognition of the omnipresence of the body of Christ as demanded by the Stuttgart church consistory. Therefore the consistory admonished him: "Do not trust your own ingenuity too much and see to it that your faith is not founded on human wisdom but on God's power."[5] As a consequence of this discord Kepler did not receive the position in Württemberg which he had desired. Galilei put the acquisition of knowledge through nature above the knowledge obtained through the Bible. He did so because he was convinced by his scientific observations that the earth is moving. When he did not change his mind in spite of several admonitions by the Church to renounce what was then considered an erroneous opinion, he was finally confined to his home and had to adjure in 1633 "his error."[6] We see here already a bifurcation of knowledge through revelation and knowledge through nature whereby knowledge through revelation is divorced from nature.

It was only in the late 17th century that theologians showed an increased interest in nature, a fact which led to the emergence of so-called physico-theology. At that point mathematization of the natural sciences had made considerable progress. René Descartes, for instance, wanted to explain the world in purely mechanical terms. Final causes should be totally excluded so that mathematical physics could be introduced as the foundational science. Yet even he could not reach the complete mechanization of nature, as becomes clear from his admission: "But to demand from me a geometric demonstration in a manner which depends on physics means to ask from me the impossible."[7]

Isaac Newton (1642–1727) succeeded for the first time in his *Mathematical Principles of the Doctrine of Nature* (*Philosophia naturalis principia mathematica*), begun in 1684 and published

---

[5] "Konsistorium in Stuttgart an Kepler in Linz," in Johannes Kepler, *Gesammelte Werke,* vol. 17: *Letters 1612–1620* (Munich: C. H. Beck, 1955), 32.

[6] Cf. Hans-Werner Schütt, "Galilei, Galileo (1564–1642)," in *TRE* 12:15.

[7] René Descartes, "Letter to Mersenne, May 27, 1638," in *Oeuvres de Descartes: Correspondence*, ed. Charles Adam and Paul Tannery, vol. 2: *March 1638 – December 1639* (Paris: J. Vrin, 1969), 142.

in 1687, to explain with a single mathematical law the phenomena of the heavens, of the tides, and the movements of objects on earth. His mathematical insights considerably helped astronomers and scientists to understand relationships in nature. Yet for Newton himself the ability to understand nature in mathematical terms revealed the greatness of God. Therefore he concluded his *Principles* with the confession:

> If the fixed stars are the centers of other like systems, these, being formed by the likewise counsel, must be all subject to the dominion of One. ... This Being governs all things, not as the soul of the world, but as Lord over all. ... It is allowed by all that the Supreme God exists necessarily; and by the same necessity he exists *always* and *everywhere.*[8]

Though Newton is known to most people as a natural scientist, most of his writings are in theology. Yet for his contemporaries these writings were not very convincing.

> To the eighteenth and much of the nineteenth centuries, Newton himself became idealized as the perfect scientist: cool, objective, and never going beyond what the facts warrant to speculative hypotheses. The *Principia* became the model of scientific knowledge, a synthesis expressing the Enlightenment conception of the universe as a rationally ordered machine governed by simple mathematical laws.[9]

For some, such as IMMANUEL KANT (1724–1804), even the fundamental principles from which the system of the world was derived seemed to be *a priori* truths, which are accessible only to reason. Yet in the natural sciences still the optimistic idea prevailed that one could simply connect the discoveries in the natural sciences with the Christian faith. It was believed that the scientific penetration of nature would even more clearly reveal the greatness of God. The Dutch scholar, theologian, and mayor

---

[8] Isaac Newton, *Mathematical Principles*, trans. Florian Cajori (New York: Greenwood, 1969), 2:544 f.

[9] Dudley Shapere, "Isaac Newton," in: *Encyclopedia of Philosophy*, 5:491.

BERNHARD NIEUWENTYT (1654 – 1718) wrote in his *Proper Use of the Understanding of the World* (1715): "From all that has been already said, it may be inferred that the exact and experimental observations of what we see in the world, is a demonstrative means, not only to obviate so many causes and inducements to *atheism,* but likewise to attain to the knowledge of a God and his perfections by his works."[10] Similarly, the Swiss natural scientist and philosopher CHARLES BONNET (1720 – 1793) concluded in his voluminous *Contemplation of Nature* (1764 – 1765):

> But this contemplation would prove fruitless, did it not lead us to aspire incessantly after this adorable Being, by endeavoring to acquire knowledge of him from the immense chain of various productions where in his power and wisdom are displayed with such distinguished truth and undiminished lustre. He does not impart to us the knowledge of himself immediately; that is not the plan he has chosen; but he has commanded the heavens and the earth to proclaim his existence, to make himself known to us. He has endued us with faculties susceptible of this divine language, and has raised up men whose sublime genius explores their beauties, and who become his interpreters.[11]

The exuberance concerning these new insights from the "book of nature" was not only a boon for theology. One could now also deepen with the help of scientific knowledge the insights concerning the existence and the essence of God gained from Scripture. When Johannes Kepler emphasized that this knowledge from nature was "of the same kind as the divine knowledge" and when the scientist was elevated to become the "priest of the highest God in the book of nature," these estimates were potentially dangerous for theology.[12] We remember that Galilei had already asserted that, when in doubt, insights derived from na-

---

[10] Bernhard Nieuwentyt, *The Religious Philosopher* or, *The Right Use of Contemplating the Works of the Creator,* trans. John Chamberlayne, 2nd ed. (London, 1721), Preface, Sect. XXVII, xlii.

[11] Charles Bonnet, *The Contemplation of Nature,* trans. from the French (London: T. Longman, 1766), 2:229.

[12] Johannes Kepler, *Gesammelte Werke,* vol. 13 (Munich: C. H. Beck, 1945), 309 and 193.

ture have priority over insights from the Bible. Yet initially physico-theology still demonstrated the unity of God and nature. The reason for this was that many theologians were active in the natural sciences. For instance the British theologian and botanist John Ray wrote in 1691 a work with the telling title *The Wisdom of God Manifested in the Works of Creation*. WILLIAM DERHAM (1657 – 1735), an English parson and philosopher of nature, calculated the speed of sound with relatively high exactitude. In 1713 he published a *Physico-theology* which did not just present physico-theological arguments but also in a compendium-like fashion presented the reader the then existent knowledge in the natural sciences. Two years later he published an *Astro-theology* in which he devoted himself to celestial objects.

In Germany, too, there was this union of theology and the natural sciences in one person, for instance in JAKOB CHRISTIAN GOTTLIEB SCHÄFFER (1718 – 1790). Schäffer was superintendent of the Protestant congregation in Regensburg but also had made himself a name as a botanist, mycologist—he published research on Bavarian and Palatine sponges—and as an entomologist. Three volumes alone with more than 3,000 drawings of insects in the Regensburg region bear witness to his scholarly interests. His collection of objects from nature was so famous that even Johann Wolfgang von Goethe came to see it in 1786. Schäffer intended to raise the living standard of the people, for instance by building a washing machine and demonstrating how paper could be produced by using plant fiber. But for his intensive research in nature there was a typically physico-theological reason: "Without excuse the first and ultimate reason for all human considerations, actions, and works should be the knowledge and veneration of the Creator and the Highest Being from his works and in them. This must also be the same with regard to insects," so the argument of Schäffer in his work *Zweifel und Schwierigkeiten welche in der Insektenlehre noch vorwalten* (*Doubts and Difficulties Which Still Prevail in the Knowledge of Insects*) of 1766.[13]

---

[13] As quoted in Markus Tanne, "Jakob Christian Schaeffer – der Superintendent als Naturforscher (1718 – 1790)," in Karlheinz Dietz/Gerhard H.

One could also mention here the Hamburg polyhistorian Jo-
HANN ALBERT FABRICIUS (1668 – 1736) who translated the works
of Derham into German and also wrote several physico-theological works. The leading thought for his scientific investigations was the "application for respectful thanks and praise to
the great Creator."[14] – Since in Königsberg there was a physico-
theological society to which, among others, young Immanuel
Kant belonged, it is no surprise that he engaged extensively with
physico-theology. With regard to recognize "God from his
works", he affirmed that the "endeavors of Derham, Nieuwentyt
and many others have ... done honor to human reason.[15] When
he mentions in conclusion: "It is absolutely necessary to convince one's self of the existence of God; but it is not just so
necessary that it should be demonstrated," he has accurately
characterized the intention of most physico-theologians.[16]

But all these different theologies which were related to nature
and developed in the course of natural scientific discoveries could
not hide the fact that God had gradually lost an appropriate place
in nature. Nature was largely understood as an arrangement of

---

Waldherr, eds., *Berühmte Regensburger: Lebensbilder aus zwei Jahrtau-
senden* (Regensburg: Universitätsverlag, 1997), 178.

[14] Johann Albert Fabricius, *Hydro-Theologie, oder Versuch, durch auf-
merksame Betrachtung der Eigenschaften, reichen Austheilung und Bewe-
gung der Wasser die Menschen zur Liebe und Bewunderung Ihres Gütigsten,
Weisesten, Mächtigsten Schöpfers zu ermuntern. [...]* (Hamburg: König and
Richter, 1734), I,2.

[15] Udo Krolzik, "Physikotheologie," in *TRE* 26:594, writes: "The most
important teacher of Kant, the philosopher Martin Knutzen (1713 – 1751)
founded at Königsberg in 1748 a physico-theological society to which besides young Kant also belonged J. G. Hamann." For the quotation Immanuel Kant, *Der einzig mögliche Beweisgrund von einer Demonstration
des Daseins Gottes* (1763), in Kant, *Werke in zehn Bänden,* Wilhelm Wei-
schedel, ed., (Darmstadt: Wissenschaftliche Buchgesellschaft , 1968), 2:734
(A 199); and the English translation *The Only Possible Argument for the
Demonstration of the Existence of God,* in Kant, *Essays and Treatises on
Moral, Political, Religious and Various Philosophical Subjects,* vol. 2 (London: William Richardson, 1799), 362.

[16] Kant, *Der einzig mögliche Beweisgrund,* 2:738 (A 205), and Kant, *The
Only Possible Argument,* 366.

geometric figures and numbers. Newton still needed God so that under the influence of gravitation the stars would not collapse and that in the face of planetary changes the stability of the solar system was maintained. Yet the progress in scientific discoveries continuously diminished the necessity for such a "God of the gaps." Very quickly physico-theology was pushed to the side by the Enlightenment. God was removed from nature and its innate laws were emphasized. The Swiss polyhistorian ALBRECHT VON HALLER (1708 – 1777) had already to defend his physico-theology against deism and skepticism being disseminated by French thinkers especially by Voltaire and La Mettrie.[17] Within less than a hundred years the situation had completely changed.

When the French mathematician and astronomer PIERRE LAPLACE (1749 – 1827) presented to Napoleon the first two volumes of his five-volume work *Mécanique céleste* (*Celestial Mechanics*, 1799 – 1825) he allegedly replied to Napoleon's question where the proper place for God was in his system: "Sir, I do not need this hypothesis." The world made sense without reference to God. Even the hypothesis of the creator seemed no longer necessary. In 1842 the German physicist JULIUS ROBERT MAYER (1814 – 1878) formulated the First Law of Thermodynamics or the Law of Conservation of Energy. This states that within an energetically isolated system the amount of energy neither increases nor decreases. This law seemed to make it possible that the world could be accorded the attribute of eternity. Given the presupposition that it is an energetically closed system it has neither a beginning nor an end. It is eternal and neither needs a creation nor a creator.

This kind of materialism was especially pronounced by the French physician and philosopher JULIEN OFFRAY DE LA METTRIE (1709 – 1751). In his book *Man as Machine* (*L'homme machine*, 1748), he presented a naturalistic view of humanity and explained spiritual processes through physiological causes. The soul, for instance, originates from the organization of the body,

---

[17] Cf. Wolfgang Wiegrebe, *Albrecht von Haller als apologetischer Physikotheologe Physikotheologie: Erkenntnis Gottes aus der Natur?* (Frankfurt am Main: Peter Lang, 2009), esp. 522 ff.

and the higher development of the reasonable human soul is due to the larger and more intricate arrangement of the brain. According to La Mettrie, this thoroughgoing naturalism necessarily leads to atheism. Already in 1745 in his *History of the Soul* (*Histoire naturelle de l'âme*) he rejected a metaphysical dualism and explained the faculties of the human mind through a motor-like power which resides in matter. The German baron PAUL HEINRICH DIETRICH VON HOLBACH (1723–1789) pursued a similar approach in his book *System of Nature* (*Système de la nature*, 1770). He described humanity as a product of nature which is subjected to the laws of the physical universe and beyond these there are no further ultimate principles or powers. According to von Holbach, it is an illusion to consider the soul as a spiritual substance. The moral and intellectual attributes of humanity can best be explained in a mechanistic way caused by physical, biological, and social interactions. The empirical and rational exploration of matter provides for von Holbach the only possibility to understand a human being. Nature is the sum total of all matter and of its movement. Matter is actually—or at least potentially—in movement, since energy or power is a property innate in matter. The material universe is simply there and therefore we need not pose the question how matter was created. Nothing is accidental or disorderly in nature since everything occurs by necessity and in an order which is determined by the universal chain of cause and effect. The world in which we live is not only interpreted in a mechanistic way, but von Holbach also believed it was subjected to a stringent causal determinism. – This was for the most part the general mood which prevailed among many intellectuals at the beginning of the 19th century. Therefore it was no accident that in November 1793, in the wake of the French Revolution, God was officially abolished and in God's place the goddess of reason was enthroned.

2. Theology in Retreat (19th Century)

From his materialistic conviction Friedrich Engels (1810–1895), the co-founder of Marxism, emphasized: "We have the certainty that matter remains eternally the same in all its transformations, that none of its attributes can ever be lost, and therefore, also, with the same iron necessity that it will exterminate on earth its highest creation, the thinking mind, it must somewhere else and at another time again produce it."[1] Nature moves in an eternal circle without beginning and end and matter is endowed with the attribute of eternity. In such a worldview there is no longer a place for theology. It came as no surprise that as a consequence theology withdrew to its innermost expertise, on the one hand in pietism to the interiority and on the other hand to a new confessionalism which expected orientation in life from Scripture and the denominational confessions. With the exception of social problems, the world was surrendered to the natural sciences. In Germany theologians were confronted with the dissemination of a materialistic and ultimately monistic worldview which was advocated in an especially influential manner by Ludwig Büchner, Carl Vogt, and Jacob Moleschott. Ludwig Büchner's book *Kraft und Stoff* (*Force and Matter*)[2] al-

---

[1] Friedrich Engels, *Dialectics of Nature* (1873–83), trans. and ed. Clemens Dutt, pref. and notes J.B.S. Haldane (New York: International Publishers, 1940), 24 f.

[2] Ludwig Büchner, *Force and Matter or Principles of the Natural Order of the Universe with a System of Morality Based Thereon: A Popular Exposi-*

ready in its title points to the two main components of materialism, force and matter. In the second half of the 19th century Büchner's book was the publication that had the widest circulation of any popular philosophical writing. In 1904 it had already gone through twenty-one editions and had been translated into all the major languages.[3]

2.1 The Materialistic Attack

Ludwig Büchner (1824–1899), a medical doctor from Darmstadt, advanced a "homogeneous theory of the world" and claimed that the "conquest of science … makes the old theistic theory of the universe, which originated in the days when mankind was still in its first childhood, appear as a mere fable."[4] He observed that there is a certain relationship between the environmental conditions and the existence of certain forms of organic life.[5] Earlier forms disappear and new ones come up as soon as the external conditions are changed.

While Büchner assumed that occasionally still today some forms of life originate spontaneously, he vacillated with regard to the origin of life itself.[6] On account of fossil findings he knew that usually lower forms of life date earlier than higher ones and that from these the ascent to further evolution occurs. When he noticed that different species resemble each other so much in their embryonic developmental state that this could only be explained by a common history of origin he anticipated what Ernst Haeckel later called the biogenetic law. Büchner rejected any thought of development by intentional planning when he declared: "Form is not a *principle* but a *result;* it is not the execution of a predesigned

---

tion, trans. from the 5th German ed., 4th Engl. ed. (London: Asher, 1884), xxv.

[3] According to Emanuel Hirsch, *Geschichte der neuern evangelischen Theologie im Zusammenhang mit den allgemeinen Bewegungen des europäischen Denkens,* 3rd ed. (Gütersloh: Gerd Mohn, 1964), 5:585 f.

[4] Büchner, *Force and Matter,* Pref. to the 15th ed., xxivf.

[5] Cf. Büchner, *Force and Matter,* 125.

[6] Cf. Büchner, *Force and Matter,* 181.

plan, but the necessary product of the interaction of a large number of causes, contingencies or energies, [even though] blind and unconscious in themselves, [continue working] everywhere and at all times without cessation, and cannot but produce an apparently perfect and graduated order and succession."[7] Büchner replaced the idea of a goal-giving force external to nature with a self-contained system. "The secret of Nature lies in an eternal and self-sustained circle, wherein cause and effect are united without beginning and without end. That which is eternal only can be from eternity, and cannot be created or made."[8] In their bodily and spiritual existence humans are a product of nature. "Man with all his eminent qualities and faculties is not a work of God but a product of Nature, like all his fellow-creatures, and has proceeded from natural and gradual evolution and self-education."[9] If nature determines everything, it is not surprising that matter is immortal, without beginning and end in space, and that force is immortal too. Büchner concludes: "Today the indestructibility or permanence of matter is a scientific fact firmly established and is no longer to be denied."[10]

It is not surprising that in Büchner's system there is hardly any space left for God. If God is eternal, so Büchner, then this is just another expression for the eternity of the world.[11] Büchner rejected especially a planned purposiveness in nature and therewith denied the idea of a creator God. In analogy to Kant, Büchner argued that the idea of teleology is derived from our own mind and cannot be deduced from nature.[12] For instance, a deer does not have its long legs to run fast, but it can run fast because it has long legs. As soon as Darwin's thesis of non-directed "mutations" became known Büchner readily accepted it to undergird his assertions. The world

---

[7] Büchner, *Force and Matter*, 91 f.

[8] Büchner, *Force and Matter*, 37.

[9] Büchner, *Force and Matter*, 248.

[10] Büchner, *Force and Matter*, 24.

[11] Cf. Ludwig Büchner, *Der Gottes-Begriff und dessen Bedeutung in der Gegenwart: Ein allgemein-verständlicher Vortrag* (Leipzig: Theodor Thomas Verlag, 1874), 18.

[12] Cf. Büchner, *Der Gottes-Begriff*, 25 f.

is no longer the best possible world as GOTTFRIED WILHELM
LEIBNIZ (1646–1716) had claimed but it is a world full of rest-
lessness and incompletion. Nevertheless, Büchner did not resign in
his advocacy of "atheism or philosophical monism" in face of
destiny. Since the motivation for our ethos no longer depends on a
transcendent God the monism of nature leads us "to freedom,
reason, and progress, to acknowledgement of humans and true
humanity – in *one word* – to *humanism.*"[13]

While the monistic understanding of the world as represented
by Büchner had an especially popular attraction others reminded
their audience that one cannot so easily dispense with Kant's
distinction between the world of phenomena and the world as a
"thing in itself."[14] Already the pessimistic philosopher ARTHUR
SCHOPENHAUER (1788–1860), who thought to be a true disciple of
Kant had maintained: "I must further remind the reader here of the
proof of the insufficiency of materialism, … because … it is in the
philosophy of the subject which forgets itself in its calculation."[15]
Can there really be an object matter without a subject and ultimately without objectivity? The neo-Kantian philosopher FRIE-
DRICH ALBERT LANGE (1828–1875) conceded in his classic *Ge-*
schichte des Materialismus und Kritik seiner Bedeutung in der
Gegenwart (*History of Materialism and Criticism of Its Significance*
in the Present; 1866) that materialism is a scientific method of
investigation.[16] He rejected, however, any metaphysical conclusions drawn from materialism since he contended that we must

---

[13] Büchner, *Der Gottes-Begriff,* 46.

[14] Cf. Julius Frauenstädt, *Der Materialismus: Seine Wahrheit und sein*
Irrtum. Eine Erwiderung auf Dr. Louis Büchners "Kraft und Stoff" (Leipzig:
F. A. Brockhaus Verlag, 1856).

[15] Arthur Schopenhauer, *The World as Will and Idea,* 3rd ed., trans. R. B.
Haldane/J. Kemp (London: Kegan Paul. Trench, Trübner, 1896), 3:59
(chap. 24).

[16] Cf. Friedrich Albert Lange, *Geschichte des Materialismus und Kritik*
seiner Bedeutung in der Gegenwart. Zweites Buch: Geschichte des Materia-
lismus seit Kant, intro. Hermann Cohen (Leipzig: J. Baedecker Verlag,
1896),538 f. In the second ed. of 1873–75, the book is divided into 1. "The
History of Materialism up to Kant" and 2. "The History of Materialism since
Kant."

confine our assertions to the world of phenomena while the things in themselves are withdrawn from our sense of experience.

As CARL VOGT (1817 – 1895) showed in *Köhlerglaube und Wissenschaft: Eine Streitschrift gegen Hofrath Rudolf Wagner in Göttingen (Backwoods Faith and Science: A Pamphlet Against Councilor Rudolf Wagner in Göttingen*; 1854) it is difficult to argue with representatives of materialistic monism. While Büchner attacked the chemist JUSTUS LIEBIG (1803 – 1873) in *Force and Matter* quite often because Liebig excluded materialistic consequences from the sciences, Carl Vogt, Professor of Zoology in Giessen and later in Geneva, vehemently attacked the Göttingen Professor of Medical Science RUDOLF WAGNER (1805 – 1864). This debate climaxed in 1852 in a literary exchange in the *Augsburger Allgemeinen Zeitung* and then in 1854 – 55 it led to the so-called "Debate on Materialism," a significant event in the intellectual and religious history of Germany. In 1854 at the 31st Congress of Natural Scientists in Göttingen this controversy reached its peak when the religiously conservative Wagner postulated the descent of all people from a single couple which at the same time corresponded to the ideal humans which he found in the Indo-European race. Though Wagner conceded that intellectual impressions and activities are conditioned by the interchange between brain and nerves he nevertheless tried to show that this did not exclude the existence of the special substance called "soul" which could neither be weighed nor made visible. Therefore the existence of individual immortal souls cannot be denied. His lecture with the title "Creation of Humans and the Substance of the Soul" was followed in the same year 1854 by another publication: *Über Wissen und Glauben mit besonderer Beziehung auf die Zukunft der Seelen (Concerning Knowledge and Faith with Special Relationship to the Future of Souls)*. Carl Vogt responded to these writings in 1854 with the publication of *Köhlerglaube und Wissenschaft (Backwoods Faith and Science)*, which subsequently was reprinted several times.

With the juxtaposition of *Köhlerglaube und Wissenschaft* Vogt showed that he was not willing to take the other side seriously. Vogt claimed that the fossils found in excavations showed that humans were not the last species which appeared on earth because many species had since then been replaced by the others. Moreover, scientific investigations had shown that the different human races could not possibly have descended from one couple. Vogt therefore concluded: "*The teachings of Scripture concerning Adam and Noah and the twofold descent of humanity from one couple are scientifically untenable fairytales.*"[17] Scientific results also showed for Vogt that human existence is only a transitory one after which no further life will follow.[18] Again, monistic materialism is seen as some kind of liberation. Since our brief stay on earth does not eventuate in "a sinister vengeance" we may live as equals and enjoy the pleasures of life.

Vogt's remark about a life "as equals" points to his socialist leanings which rejected, as we can also detect with Büchner, the authoritative faith, the monarchy, and the rule of the priest.[19] The Dutch physiologist JACOB MOLESCHOTT (1822 – 1893) was even more blatant about the relationship between science, which meant for him materialism, and socialism when he said: "The naturalist is the most engaged advocate of the social issue. … Its solution lies in the hand of the naturalist which is guided with certainty by the experience of the senses."[20] As we have also seen with Friedrich Engels materialism in the 19th century was atheistic and, quite different from today, at the same time revolutionary in social matters.

If we now turn briefly to Jacob Moleschott and his book *Der Kreislauf des Lebens* (*The Circle of Life*; 1852) we notice that he radicalized most of the issues introduced by Büchner and Vogt. It is not surprising that Büchner confessed that Moleschott had inspired him to write *Kraft und Stoff*. The sequence of cause and effect were so determining for Moleschott that he could not admit any ultimate dependence in contrast to what Schleiermacher had

---

[17] Carl Vogt, *Köhlerglaube und Wissenschaft: Eine Streitschrift gegen Hofrat Wagner in Göttingen*, 4th ed. (Gießen: J. Ricker, 1856), 83.

[18] Vogt, *Köhlerglaube*, 122.

[19] Büchner, *Der Gottes-Begriff*, 45.

[20] Jacob Moleschott, *Der Kreislauf des Lebens: Physiologische Antworten auf Liebigs Chemische Briefe*, 2nd ed. (Mainz: Victor von Zabern, 1855), 480.

still advanced. "Research excludes revelation."[21] The omnipotence of the Creator of the world stands in irreconcilable opposition to the laws of nature. Revelation and knowledge are related to each other like fiction and truth; the one guesses where the other researches. Truth can only be gained from nature and its course. The essence of a thing is the sum total of its properties and the essence of all properties is force. But force is no divine means, no essence of the things separate from its foundation; it is rather the inexorable eternal inner property of matter. For both Moleschott and Büchner force and matter were identical. There is only one substance which is immortal, namely matter. The unchangeableness of matter is the foundation for the eternal circle in nature of becoming and perishing.[22]

The immortality of matter Moleschott found attested to by JULIUS ROBERT MAYER'S (1814–1878) Law of the Conservation of Energy, which says that the amount of energy in an energetically closed system remains constant.[23] Moleschott was certainly right when he claimed that by transforming energy from one form into another no energy gets lost. We can also agree with him that energy is as indestructible as matter. Yet it is not insignificant that Moleschott passed over the Law of Entropy, the second law of thermodynamics, which means the law of the inconvertibility of energy which was advanced in 1851 by the physicist RUDOLF EMANUEL CLAUSIUS (1822–1888). Therefore Moleschott could still claim: "The supply of force in the universe is always the same" and "that the sum of the tension and of the living force in the universe remain always the same."[24] Energy is indestructible and guaranties the unwavering course of an eternal process.

While Büchner was still unsure whether the first organism originated through a spontaneous genesis, this issue was resolved

---

[21] *Moleschott, Der Kreislauf des Lebens*, 2nd ed., 18; Moleschott, *Der Kreislauf des Lebens,* 5th ed., vol. 1 (Gießen: Emil Roth, 1875), 7.

[22] Cf. Moleschott, *Der Kreislauf des Lebens,* 5th ed., vol. 1, 35.

[23] Cf. Moleschott, *Der Kreislauf des Lebens*, vol. 2 (Gießen: Emil Roth, 1887), 608 f.

[24] Moleschott, *Der Kreislauf des Lebens*, vol. 2, 609.

for Moleschott in 1887.[25] He was certain that the claim that life on earth had originated from lifeless matter could never be contradicted even if we could never produce a spontaneous genesis by experiments. Present-day humanity is part of the evolutionary process and has arisen through the primal cell, the monkeys, and primates. Presumably the whole evolutionary process had started with inorganic matter and developed up to humanity. Therefore Moleschott interpreted humanity in strictly materialistic terms. "Without phosphorous, without fat, without water, there is no thought," was Moleschott's judgment.[26] "Speech and style, trials and conclusions, benefits and crimes, courage and half-heartedness and treason, these are all natural events which all stand as necessary consequences in a strict relationship to indispensable causes."[27] Humans are the end product of parents, space and time, air and weather, food and clothing. Since humanity is exposed to so many influences it is a continuously further developing product of nature.[28]

Nevertheless, Moleschott breaks through the thoroughgoing determinism he had constructed. While all of our knowledge comes from our senses and while the history of human education is the history of the development of our senses, he declares: "Humans are the measure of all things for humans."[29] Since we are continuously becoming we can continuously improve ourselves. Actually, he declares, "the ethical and mental faculties of the human race are in the process of continuous growth."[30] In his appeal to humanity to realize the new possibilities of a monistic materialism Moleschott implicitly went beyond a strict monistic view and distinguished between one's own self and the world. In spite of their strained assertions to the opposite, this relationship between one's own self and matter is exactly the problem that

---

[25] Cf. for the following Moleschott, *Der Kreislauf des Lebens*, vol. 2, 593.

[26] Moleschott, *Der Kreislauf des Lebens*, vol. 2, 599.

[27] Moleschott, *Der Kreislauf des Lebens*, vol. 2, 606.

[28] Cf. Moleschott, *Der Kreislauf des Lebens*, vol. 2, 608.

[29] Moleschott, *Der Kreislauf des Lebens*, vol. 2, 583.

[30] Moleschott, *Der Kreislauf des Lebens*, vol. 2, 613.

pre-Darwinistic materialistic monists evidently could not solve when they attempted to put forth a coherent worldview.[31]

Though we do not want to regard LUDWIG FEUERBACH (1804–1872) as a materialistic monist we will mention him at least briefly in this context. Feuerbach was more concerned with philosophy of religion than with philosophy of nature. Nevertheless his concepts were strongly influenced by evolutionary thought. In his treatise *The Essence of Religion* (1851) he claimed that religion arises from the feeling of dependence.[32] God is the highest and most powerful being and can accomplish what humans cannot. God is the eternal and also the first cause who serves as a hypothesis to solve the issue of the origin of nature, respectively of organic life. It is the power of God which sustains us. But then Feuerbach asks: "We are placed right in the midst of Nature and should our beginning, our origin, lie outside of Nature? We live within Nature, with Nature, by Nature, and should we still not be of her? What a contradiction!"[33] Indeed, Feuerbach concluded, that the basic concept of God is nothing but nature in contrast to humanity.

Different from his teacher GEORG WILHELM FRIEDRICH HEGEL (1770–1831) Feuerbach declared that the deduction of the world from God or nature from spirit is nothing but a game in logic. Religion therefore has as its presupposition the dichotomy or even the contradiction between intention and ability, desire and fulfillment, idea and reality. The traditional understanding of faith in God is an anthropomorphic phenomenon of faith in nature. Feuerbach, however, wanted to turn this faith into a faith in humanity who is in and from nature.[34] In trying to merge God into nature Feuerbach did not intend a divinization of nature. To the contrary, Feuerbach argued: "The power of Nature is not unlimited like the power of God, *i. e.* the power of human imag-

---

[31] Cf. Theobald Ziegler, *Die geistigen und sozialen Strömungen des neunzehnten Jahrhunderts*, 2nd ed. (Berlin: Georg Bondi, 1901), 350.

[32] Cf. Ludwig Feuerbach, *The Essence of Religion*, trans. Alexander Loos (Amherst, NY: Prometheus Books, 2004), 1 (§ 2).

[33] Feuerbach, *The Essence of Religion*, 16 (§ 16).

[34] Cf. Feuerbach, *The Essence of Religion*, 64 (§ 53).

ination; she cannot do everything at all times and under all circumstances – her productions and effects on the contrary are dependent on conditions."[35] This thesis was for him the starting point of his evolutionary ideas. Geology had proven that our earth went through a series of developments to attain its present state. Origination and development of organic life are closely connected with the evolution of the earth. Yet the earth had not first provided the conditions for life or respectively the human species as some kind of Garden of Eden in which then life or the human species originated.

As soon as the conditions on earth existed for the origin of life or of humanity then, Feuerbach concluded, life and humans originated. Therefore the assumption is wrong that life in the past could not have originated from inorganic matter just because today there is no spontaneous genesis. The environmental conditions of today are different and the earth today is in a state of stability. Yet this did not mean for Feuerbach that the evolutionary energy had been lost. His explanation in which he attempted to equate God with nature was a thesis leading to further evolutionary development. If humanity would no longer be present for nature, nature could still be present for humanity; humanity can therefore attain self-determination, dignity, and power.[36] – If we consider that Feuerbach wrote in the Preface to Jacob Moleschott's *Lehre der Nahrungsmittel: Für das Volk* (*Teaching About Food: For the People*; 1850) that "Humans are what they eat," then we could rightly doubt whether Feuerbach succeeded better than the monistic materialists to determine the relationship between one's own self and the world.

---

[35] Feuerbach, *The Essence of Religion*, 17 (§ 18).

[36] Feuerbach, *The Essence of Religion*, 42 (§ 39): "There Nature is the object of adoration, here of enjoyment, there man exists for Nature's sake, here Nature for man's sake, there she is the end, here the means; there she stands above, here below man."

2.2 The Evolutionistic Attack

Until the middle of the 19th century the discussion in the history of thought was largely dominated by materialistic ideas. But in the second half of the 19th century Charles Darwin and his book *The Origin of Species* (1859) signaled a second offensive movement against the Christian faith. The whole 19th century was governed by an immense faith in progress. Yet the scientific discoveries of Darwin, that humans would not only develop further but can be traced back to pre-human life forms, shook the faith in the special place of humanity within creation. Theologians had neglected that the scientists had been thinking in evolutionary terms for a long time. For instance, HEINRICH FRIEDRICH LINK (1767–1851), who taught as professor of natural history at the University of Berlin from 1815 onward, claimed in his book *Die Umwelt und das Altertum, erläutert durch die Naturkunde* (*The Environment and Antiquity Explained Through the Knowledge of Nature*; Berlin: 1820–22), that historically speaking monkeys are the connecting link between animals and humans. He was also convinced that Africans are the original form of the human race while whites are a degeneration thereof.[37]

OTTO ZÖCKLER (1833–1906), a theologian who taught at the University of Greifswald, even pointed out that already since the middle of the 1840s each year several scientific monographs and even textbooks had been published in various areas which anticipated Darwinian theories.[38] The ground was so well prepared that as soon as essential ideas of Darwin were made known many of his German followers soon surpassed him with their conclusions which they gathered from evolutionary ideas. In their enthusiasm they did not even refrain from crass scientific exaggerations and distortions. Carl Vogt, for instance claimed that humans who were born with abnormally small brains were an

---

[37] So Otto Zöckler, *Geschichte der Beziehungen zwischen Theologie und Naturwissenschaft mit besonderer Berücksichtigung der Schöpfungsgeschichte*, vol. 2: *Von Newton und Leibniz bis zur Gegenwart* (Gütersloh: C. Bertelsmann, 1879), 604 f.

[38] Cf. Zöckler, *Geschichte der Beziehungen*, 613.

essential link between humans and monkeys. They are examples
of a pathological malformation which leads back to the primates
as our ancestors. Under heavy attack Carl Vogt had to withdraw
his theory and admitted in 1872 that he had never investigated
such a microcephalic brain. There were also some moderate re-
ceptions of Darwin's theories. But the most rigorous and most
enduring influence of Darwinian ideas in Germany though
continuing pre-Darwinian monistic tendencies were exerted by
David Friedrich Strauss and Ernst Haeckel.

At first glance it may be improper to begin our survey of
monistic tendencies in German Darwinism with DAVID FRIE-
DRICH STRAUSS (1808–1874). He was a representative of Hegelian philosophy and not so much of Darwinian thoughts. In his
*Christliche Glaubenslehre in ihrer geschichtlichen Entwicklung
und im Kampfe mit der modernen Wissenschaft dargestellt*
(*Christian Doctrine of Faith, Its Historic Development and Conflict with Modern Science*; vol. 1, 1840) he employs a strict Hegelian methodology by dividing each part in thesis, antithesis and
synthesis (616).[39] He agreed with Hegel that the Absolute is essentially a result; only at the end the total reality appears (643).
Evolutionary or progressive thinking corresponds exactly with
the center of his understanding of the relationship between God
and creation. Strauss was convinced that all of organic life originated in subsequent layers according to evolution from lifeless
matter (681 f.). Only gradually has our planet gained its present
shape. There has been a creative power at work on our planet
which preserved the created and mediated the preservation of
higher forms of life through procreation (685 f.). Strauss however
pointed out that neither the polygenetic origin of humanity nor
the naturalness of its origin mitigates against the thought of the
unity of the human race and the idea of a creator God. While
Strauss did not yet know details of the evolutionary process he
was convinced of a natural development of creation.

---

[39] The parenthetical page numbers in this paragraph are from Strauss,
*Die christliche Glaubenslehre in ihrer geschichtlichen Entwicklung und im
Kampfe mit der modernen Wissenschaft dargestellt,* vol. 1 (Darmstadt:
Wissenschaftliche Buchgesellschaft, 1973 [1840]).

When we turn to *The Old Faith and the New* (1871; Eng. trans. of the 6th ed. 1873), which Strauss published one year after Darwin's *Descent of Man* we cannot but realize the amazing impact the Darwinian theory had on him. Gone now is the possibility of reconciling the progression of nature with the belief in a Creator. We are taught that the only choice is between belief in the divine creative hand and belief in Darwin's theory. "Natural Science," Strauss declared, "has long endeavored to substitute the evolutionary theory in place of the conception of creation so alien to her spirit; but it was Charles Darwin who made the first truly scientific attempt to deal seriously with this conception, and to trace it throughout the organic world."[40] Of course, he admitted that Darwin was not the first to suggest this theory. Most other proponents of this theory, however, had too many parts missing in it to introduce it as a convincing and comprehensive theory. While Strauss conceded that Darwin pointed toward more possible solutions than he actually provided, he was convinced that Darwin had introduced the evolutionary process so persuasively that "a happier coming race will finally cast out miracles" from creation (1:205). Thus Strauss declared Darwin "one of the greatest benefactors of the human race" (1:205). For the progressive natural science of today an intelligent architect of the organism, or even a purposiveness of nature that one could understand as the work of an intelligent creator, was no longer tenable (cf. 2:23 – 25). Even human instincts were gradually acquired through natural selection. This showed, for Strauss, the immense chasm between the old worldview and the new one.

Strauss did not think that the details of Darwin's theory were surprising. He argued, for instance, that "Darwin's 'struggle for existence' is nothing but the expansion of that into a law of Nature, which we have long since recognized as a law of one's social and industrial life … competition" (1:216 – 17). Of course, Strauss realized that some people might find humanity's descent from apes offensive (cf. 2:4). But he questioned whether it would

---

[40] David Friedrich Strauss, *The Old Faith & the New,* 2 vols. in 1, with introduction and notes by G. A. Wells (Amherst, NY: Prometheus, 1997), 1:202 (for the following references page numbers are given in the text).

be any better to be created in the image of God and then be
thrown out of paradise, and still today not have regained the
status we once had. For some people, he said, even a failure from a
good family is more respectable than someone who, through his
or her own efforts, has made it in life. According to the evolutionary theory, humanity did not start high to fall so far immediately afterward. On the contrary, it started low, to move slowly
but "to ever greater heights" (2:39). Humanity is part of the as-
cending movement of life (cf. 2:55). In it nature reflects itself,
Strauss declared in agreement with his mentor Hegel. Though
humans are still natural beings, they sublimated the higher goal
implanted in them. Humanity should understand and dominate
nature, not as a tyrant but as humans. While these assertions
about the human "destiny" may be interpreted as resembling
teleological thought, it was evident for Strauss that the evolutionary theory is in opposition to any "dualistic" Christian understanding of the world. This theory endeavors to explain the
whole of phenomena from one single, monistic principle. Compared with Christianity, both materialism and idealism may "be
regarded as Monism" (2:19). Their common enemy is dualism,
"which pervaded the conception of the world throughout the
Christian era" (2:19). While idealism attempts to explain the
world from above, materialism attempts to explain it from below,
and ultimately one leads to the other. We are not wrong to conclude that Darwin's theory of development of the species through
natural selection brought for Strauss the solution to the greatest
enigmas of the world.

When we consider the approach of ERNST HAECKEL (1834–
1919) professor of zoology at the University of Jena, we are
dealing with the person who in regard to evolutionary thought
had perhaps the deepest impact on German Protestantism. Different from Büchner, Vogt, and Moleschott, he was a scholar in
his own right who made his own significant contribution to the
theory of evolution. His popular books, the two-volume *History
of Creation* (1868; Eng. trans. 1906) and *The Riddle of the Universe* (1899; Eng. trans. 1900), were bestsellers on the German
market and were translated into many languages. About his
History of Creation Charles Darwin remarked in his *Descent of*

Man: "If this work had appeared before my essay had been written, I should probably never have completed it. Almost all the conclusions at which I have arrived I find confirmed by this naturalist, whose knowledge on many points is much fuller than mine."[41] Most of Haeckel's thoughts, however, can be traced back to earlier thinkers such as Giordano Bruno, BARUCH SPINOZA (1632 – 1677), Gottfried Wilhelm Leibniz, Feuerbach, and others. Haeckel's monistic worldview is nothing new, but under the impact of the theory of a uniform evolution, it gained in precision and persuasion.

Already in his two-volume work *The History of Creation* Haeckel ended on a highly optimistic note. He found that the mind of the whole human race had gone through a process of slow, gradual, and historical development. "We are proud," he claimed, "of having so immensely outstripped our lower animal ancestors, and derived from it the consoling assurance that in the future also mankind, as a whole, will follow the glorious career of progressive development, and attain a still higher degree of mental perfection."[42] Especially the application of evolutionary thought to practical human life, as begun by the British philosopher and popularizer of Darwin's ideas, HERBERT SPENCER (1820 – 1903) opens up "a new road towards moral perfection."[43] Shaping politics, morals, and the principles of justice in accordance with natural laws, and this means with the laws of evolution, will provide "*an existence worthy of man*, which has been talked of for thousands of years."[44] The strong religious overtones in his scientific and philosophic work are most clearly expressed when he concludes:

---

[41] Charles Darwin, *The Descent of Man,* in *The Works of Charles Darwin,* ed. Paul H. Barrett/R. B. Freeman (London: William Pickering, 1989), 21:5 [original pagination: 3/4]. Darwin also mentioned approvingly the evolutionary thoughts of Büchner and Vogt.

[42] Ernst Haeckel, *The History of Creation: On the Development of the Earth and Its Inhabitants by the Action of Natural Causes,* trans. E. R. Lankester, 2 vols. (New York: D. Appleton, 1876), 2:367.

[43] Haeckel, *The History of Creation,* 2:367.

[44] Haeckel, *The History of Creation,* 2:368.

> The simple religion of Nature, which grows from a true knowledge of Her and of Her inexhaustible store of revelations, will in the future ennoble and perfect the development of mankind far beyond the degree which can possibly be attained under the influence of multifarious religions of the churches of the various nations, – religions resting on a blind belief in the vague secrets and mythical revelations of a sacerdotal caste.[45]

The religious overtones of his worldview can be seen throughout his widely publicized lecture *Monism as Connecting Religion and Science: The Confession of Faith of a Man of Science* (1893; Eng. trans. 1894). Haeckel claimed as an "article of faith" the fundamental unity of organic and inorganic nature and thereby rejected the distinctions between natural science and humanities.[46] While most older religions and philosophical systems are dualistic, distinguishing between God and world, creator and creation, spirit and matter, Haeckel advocated a uniform understanding of all nature. Reminiscent of Feuerbach's position, he contended that in dualistic systems the most fundamental thought is an anthropomorphism. This means humanity devises an anthropomorphous concept of God that is "separated by a great gulf from the rest of nature" (14). Yet in referring both to Robert Mayer's law of the conservation of energy and to Lavoisier's law of the conservation of matter, and claiming that these two laws merge into the laws of the conservation of substance, Haeckel advocated a monistic worldview. The world is made up of "the inert heavy mass as material of creation" and "the mobile cosmic ether as creating divinity" (25). Thus with the ether theory, which at that time was still upheld as the means of the propagation of electromagnetic waves, no God as creator was necessary. God is not to be placed over against the material world as an external being, but must be placed as a "divine power" or "moving spirit" with the cosmos itself (15).

---

[45] Haeckel, *The History of Creation*, 2:369.

[46] Ernst Haeckel, *Monism as Connecting Religion and Science: The Confession of Faith of a Man of Science,* trans. J. Gilchrist (London: Adam and Charles Black, 1894), 93. Page references to this work are supplied in the text.

The evolutionary theory shows that in the cosmos everything is in flux. Paleontology, comparative anatomy, and ontogeny show us how life developed step by step, and how the cosmos emerged from a chaotic primeval state to the present world order. Haeckel conceded that one could label this worldview materialism, but he felt the term "monism" was more appropriate. He did not consider his worldview atheistic in the strictest sense, though it was a-theistic, because it did not want to reduce God to "a gaseous vertebrate" as did those who hold on to "God as a 'spirit' in human form" (115). Therefore, Haeckel concluded his "monistic Confession of Faith with the words: 'May God, the Spirit of the Good, the Beautiful, and the True, be with us!'" (89). Small wonder that Haeckel sought support from Bruno and Spinoza, claiming that "of the various systems of pantheism, which for [a] long [time] have given expression more or less clearly to the monistic conception of God, the most perfect is certainly that of Spinoza" (79). Haeckel was so certain of the persuasiveness of his monistic system that he contended that if Kant had developed his philosophy now with all our scientific knowledge at hand, his "system of critical philosophy would have turned out quite otherwise from what it was, and purely monistic" (102 n. 8).[47]

In his most influential book, *The Riddle of the Universe,* Haeckel's earlier unbridled optimism has been toned down. He confessed at the outset that he is "wholly a child of the nineteenth century" and that his "own command of the various branches of science is uneven and defective."[48] But he did not desist from presenting a monistic worldview, a view of a universe that is infinite and eternal. There is a universal movement of substance in space, we hear, that "takes the form of an eternal cycle or of a

---

[47] It should be mentioned parenthetically at least that Haeckel praised Strauss greatly because he "had already clearly perceived that the soul-activities of man, and therefore also his consciousness, as functions of the central nervous system, all spring from a common source, and, from a monistic point of view, come under the same category [i.e., monism]" (Haeckel, *Monism,* 46).

[48] Ernst Haeckel, *The Riddle of the Universe at the Close of the Nineteenth Century,* trans. J. McCabe (New York: Harper, 1900), ix. Page references to this work are supplied in the text.

periodic process of evolution" (243). The eternal drama of periodic decay and rebirth of cosmic bodies does not stop short of
our own solar system. Unlike the monistic pre-Darwinian materialists, Haeckel did not shy away from discussing the entropy
or non-convertibility of energy to strengthen his point of the
eternal movement of the universe. While in any finite system
"every attempt to make such a *perpetuum mobile* must necessarily fail . . . the case is different, however, when we turn to the
world at large, the boundless universe that is in eternal movement" (246). Since the universe is infinite, the law of entropy does
not affect it; there is no beginning and no end to the universe.
Again Haeckel praised Spinoza's monistic system, so closely resembling his own (290).

As once before expressed in *Monism as Connecting Religion and
Science,* Haeckel had a twofold objective: (1) "to give expression to
the rational view of the world which is being forced upon us with
such logical rigor by the modern advancements in our knowledge
of nature as a unity," and (2) to make monism into a connecting
link "between religion and science, and thus contribute to the
adjustment of the antithesis so needlessly maintained between
these, the two highest spheres in which the mind of man can ex-
ercise itself."[49] It is interesting that in so doing he also spoke out
against Christian ethics, since it is "a very ideal precept, but as
useless in practice as it is unnatural" (353). On principle it attacks
and despises egotism and exaggerates the love of one's neighbor at
the expense of self-love. We are reminded of Friedrich Nietzsche
(1844 – 1900), who went even further on this point, labeling
Christian ethics as slave-morality, which he contrasted with a
master-morality *(Herren-Moral).*[50] Yet the greatest surprise awaits
us at the conclusion of *The Riddle* when Haeckel confessed that
"only one comprehensive riddle of the universe now remains —

---

[49] Haeckel, *Monism,* vi – vii, and cf. Haeckel, *Riddle of the Universe,* 332.

[50] Friedrich Nietzsche, *Beyond Good and Evil,* trans. H. Zimmern,
(EBook, 2009), chapter IX, 260. J. B. Müller, "Herrenmoral," in *Historisches
Wörterbuch des Philosophie,* ed. Joachim Ritter/Karlfried Gründer, 3:1078,
states that Nietzsche with his theory of morals of a master race is "one of the
leading representatives of Social Darwinism."

the problem of substance" (380). Then he conceded: "We grant at once that the innermost character of nature is just as little understood by us as it was by Anaximander and Empedocles twenty-four hundred years ago, by Spinoza and Newton two hundred years ago, and by Kant and Goethe one hundred years ago. We must even grant that this essence of substance becomes more mysterious and enigmatic the deeper we penetrate into the knowledge of its attributes, matter and energy, and the more thoroughly we study its countless phenomenal forms and their evolutions" (380). But Haeckel did not want to end in resignation just because he did not discern the "thing-in-itself" that lies behind the knowable phenomena. Instead he asked us to rejoice in the immense progress that has actually been made by the monistic philosophy of nature. The monism of the cosmos "proclaims the absolute dominion of 'the great eternal iron laws' throughout the universe" (381). Instead of being devoted to the ideals of God, freedom, and immortality, we can engage in the cult of the true, the good, and the beautiful, "which is the heart of our new monistic religion" (382). Haeckel concluded that he hoped that in the twentieth century the great antithesis between theism and pantheism, vitalism and mechanism can be resolved "by the construction of a system of pure monism" (383).

The Riddle of the Universe stands at the close of the nineteenth century and is indicative of the general outlook of nineteenth-century Protestant thought in Germany. With the church historian Friedrich Loofs (1858 – 1928), one could attack the accuracy of some of the scientific arguments that Haeckel used. One can also argue that his monistic worldview is nothing new, that it is common to most philosophers in their endeavor to provide a comprehensive worldview.[51] With the pathologist and politician RUDOLF VIRCHOW (1821 – 1902), *The Freedom of Science in the Modern State,* one could also caution that all our knowledge is only partial.[52] Virchow suggested that we confine our hypotheses

---

[51] Oliver Lodge, *Life and Matter: A Criticism of Professor Haeckel's "Riddle of the Universe"* (New York: Putnam, 1907), 7, 9.

[52] Rudolf Virchow, *Die Freiheit der Wissenschaft im modernen Staat.*

to the fields for which they were designed and not expand them to universal principles (18 – 20). Against a universal theory of descent Virchow claimed that nobody had observed a spontaneous generation of life and that every progress in prehistoric anthropology has only widened the gap between human ancestors and other vertebrates (29). Therefore, Virchow concluded: *"We cannot teach, we cannot label it as an achievement of science that man is descended from apes or from any other animal"* (31). Against these objections we must note that Haeckel did not claim originality for his ideas, nor that he had all the answers, nor that his system was without flaws. As we have seen with the earlier materialistic monists, even without the detailed theory of descent Haeckel would perhaps not have wanted to abandon his thoroughgoing monism. – Similar to Zöckler, Virchow noted a close and dangerous relationship between the theory of descent and socialism (12).[53] The idea of an upward-slanted continuous evolutionary process provided the main stimulus for socialist reform claims. We should also remember that both Karl Marx and Friedrich Engels knew Darwin's work and thought that it provided the basis for their theory on the history of nature.[54] If nature and humanity have progressed so far, then the optimistic worldview of a socialist revolution would be justified.

2.3 The Reaction of Theology

How did theology respond to that challenge? Two examples may suffice. The first, Lutheran theologian CHRISTOPH ERNST

Rede (Berlin: Wiegandt, Hempel & Parey, 1877; Eng. trans. 1878), 13. Page numbers to this work are placed in the text.

[53] For further details about the dispute between Haeckel and Virchow, especially pertaining to the relationship between socialism and the theory of descent, cf. the excellent book by Ernst Benz, *Evolution and Christian Hope: Man's Concept of the Future, from the Early Fathers to Teilhard de Chardin,* trans. H. G. Frank (Garden City, N.Y.: Doubleday, Anchor Book, 1968), 96 – 98. The idea of an upward-slanting continuing evolutionary process provided the main stimulus for socialist reform claims.

[54] Cf. Benz, *Evolution and Christian Hope,* 83 ff.

LUTHARDT (1823 – 1902), professor of systematic theology and New Testament exegesis at Leipzig, beginning in 1856 emphasized in his *Die christliche Glaubenslehre gemeinverständlich dargestellt (Exposition of the Christian Faith)* that the Bible is not a book about scientific research nor the scientific knowledge of nature, but a book of religion that has to do with humanity's relationship to God and with the relationships between people on earth.[55] Referring to the materialism debate that arose in 1852, Luthardt distinguished between a psychological materialism that denies the existence of the soul and a later cosmological materialism "which denies the existence of the absolute Spirit and reduces everything existing to matter only."[56] Luthardt attributed this kind of materialistic thinking to pantheism, because it explained being from the concept of the idea. The absolute idea is everything, it reproduces being and permeates it. The Spirit then posits matter. Luthardt rejected this Hegelian dialectic. He noted that if all life is the movement of matter and all development is the development of matter, then there is no higher Spirit possible which sets a goal and an aim for the developmental process. This means that theology would be obsolete, and all that would be left is a causality with its cause-and-effect sequence.

This kind of pantheism actually becomes a monism and a materialistic pantheism. The diversity of the world is nothing but the progressive development of the world's actual foundations and beginnings. Darwinism fits well into this kind of thought pattern. Yet Luthardt asked how progress and development are possible if the effect cannot contain more than is contained in its cause. The world cannot be its own creator. If one dissociates God from the material realm, there is no spiritual world possible, not

---

[55] Cf. Christoph Ernst Luthardt, *Die christliche Glaubenslehre gemeinverständlich dargestellt,* 2nd ed. (Leipzig: Dörffling & Franke, 1906 [1898]), 212.

[56] Christoph Ernst Luthardt, *Die modernen Weltanschauungen und ihre praktischen Konsequenzen: Vorträge über Fragen der Gegenwart aus Kirche, Schule, Staat und Gesellschaft im Winter 1880 zu Leipzig gehalten* (Leipzig: Dörffling & Franke, 1880), 167, in his lecture "Der Materialismus und seine Konsequenzen."

even a humanity. Luthardt concluded: "The result of the development is either pessimism or Christendom."[57] Either the world
makes no sense whatsoever and one must despair, or one assumes
some kind of spiritual meaning or beliefs. Materialism too is a
faith; it is not a fact.

The Lutheran theologian OTTO ZÖCKLER (1833–1906), from
1886 until his death, professor of historical and exegetical theology
at the University of Greifswald, Germany, was less confrontational
than Luthardt. He attempted to demonstrate that the principles of
good science and good theology do not necessarily conflict. In his
article on Zöckler, Victor Schultze rightly claimed that his scholarship "was rated very high as was his authority as a theologian in the
realm of natural science."[58] Zöckler, who taught in Greifswald most
of his life, had indeed an astounding command of scientific
knowledge, especially in its historic dimension.

In his two volume *Geschichte der Beziehungen zwischen Theologie und Naturwissenschaft mit besonderer Rücksicht auf Schöpfungsgeschichte* (*The History of the Relationship between Theology
and the Natural Sciences with Special Attention to the Creation
Narrative*), he showed his immense erudition in the history of
science and his critical awareness. For instance, while he knew that
Darwin rejected a Christian teleological worldview, he realized that
Darwin "does not regard the sequence of the main events in the life
of nature and humanity as 'the result of a blind accident.'"[59]
Zöckler concluded that Darwin's teachings do not contain anything that would necessitate the abandonment of the Christian
theistic notion of creation.[60] While Zöckler rejected Darwinism as
a pathological disease that eventually will run its course, he was
convinced that, true to the Pauline saying that "all things are
yours" (1 Cor. 3:21), the theological doctrines of creation and
providence and the understanding of humanity's original state can

---

[57] Luthardt, *Die modernen Weltanschauungen,* 185.

[58] Victor Schultze, "Zöckler, Otto," in *The New Schaff-Herzog Encyclopedia of Religious Knowledge* (Grand Rapids: Baker, 1949–50), 12:520.

[59] Zöckler, *Geschichte der Beziehungen,* 642 f.

[60] Zöckler, *Geschichte der Beziehungen,* 719.

gain new insights from the findings of Darwin.[61] Zöckler introduced the concept of a 'theory of concordance' through which
the findings of evolutionary speculation, insofar as they are scientifically proven, complement the assertions of theology.

The hypothesis of concordance or of harmonizing is also
employed in his lectures, *Die Urgeschichte der Erde und des
Menschen (The Primal History of the Earth and Humanity)*. He
claimed that the contempt with which some scientists treat the
first chapters of the Bible can only be the result of misinformation
about the biblical Christian worldview as it pertains to the creation of the universe.[62] The eternal and infinitely grand perception contained in these stories has room for all the scientific
details through which sober empirical research will enrich our
understanding of how creation has occurred and still does occur.
Even concerning the six days of creation, there is a concordance
between the record of geology and the book of Genesis.[63] This
mutually complementing avenue had already been pursued by
the French naturalist Georges Cuvier at the beginning of the
nineteenth century, and according to Zöckler there seemed to
appear "an ever stronger consensus of all scientists in this area
even in Germany which will soon lead to a complete victory over
any contrary perspective."[64]

Zöckler's optimism was based on his historical research. He
realized that the claim made by the "modern fanatics of unbelief"
that natural science sooner or later will do away with religion and
with the Christian faith was simply not true.[65] There is no correlation between a comprehensive scientific education and religious unbelief. In each epoch there have been conservative and
decidedly irreligious scientists, but most have pursued a middle

---

[61] Zöckler, *Geschichte der Beziehungen*, 798 ff.

[62] Otto Zöckler, *Die Urgeschichte der Erde und des Menschen: Vorträge
gehalten zu Hamburg im März 1868* (Gütersloh: C. Bertelsmann, 1868), 1 f.

[63] Cf. Zöckler, *Urgeschichte*, 42.

[64] Zöckler, *Urgeschichte*, 48 f.

[65] Otto Zöckler, *Gottes Zeugen im Reich der Natur: Biographien und
Bekenntnisse großer Naturforscher aus alter und neuer Zeit*, 2nd ed. (Gütersloh: C. Bertelsmann, 1906), 482.

course. He was convinced that the future does not belong to materialism but to a true empiricism that collects and analyzes the experiences available within the realm of the visible. *"True scientists will time and again be able to read from the two texts placed alongside each other, from the Book of Nature and from the Book of Revelation. They will over and over again return to the religion of Kepler and Galilei, of Haller and Euler, and of Cuvier and Agassiz."*[66] Of course, Zöckler conceded that in the future some will become so radical as to deny the existence of everything that is not visible and tangible. But the true representatives of science will overcome these destructive forces.

True witnesses to God in the realm of nature will never die out as long as nature remains, because it is God's. Therefore human witness to the divine truth and the grandeur contained in it will never be wanting. That nature witnesses to God is especially emphasized in Zöckler's earlier publication, *Theologia naturalis (Natural Theology)*. According to the maxim *credo ut intelligam* (I believe in order to understand), he wanted to "explain, complete, and confirm the immediate revelation of God through the mediated one which is given in nature."[67] The book of nature will illustrate the Bible, while the latter will explain the former. According to Zöckler, such a positive theology of nature will expand and illustrate the organic development of dogmatics. How nature and the Bible come together can be seen especially well in the metaphors and parables of the Old and New Testaments.[68] The biblical symbols representing nature as well as the Bible's picturesque language exemplify the illustrative character of nature for God's revelation. Theological insight and scientific research do not go in separate or opposite ways but complement and in

---

[66] Zöckler, *Gottes Zeugen*, 485.

[67] Otto Zöckler, *Theologia naturalis: Entwurf einer systematischen Naturtheologie vom offenbarungsgläubigen Standpunkte aus*, vol. 1: *Die Prolegomena und die specielle Theologie enthaltend* (Frankfurt am Main and Erlangen: Heyder & Zimmer, 1860), 6. Volume 2 never appeared. This may perhaps serve as an indication that his project was more difficult than he had initially envisioned.

[68] Zöckler, *Die Prolegomena*, 200 f.

some respects even correct each other. Revelation and God's action do not occur in a realm removed from the natural world but in the midst of nature. Nature is fundamentally the arena and medium of God's action. Otto Zöckler impresses with both his historical and scientific erudition and the ease with which scientific knowledge, for him, complements theological insights. He might have been overly optimistic in asserting that the two are complementary, but at least he saw a vigorous engagement with the sciences as indispensable for theological assertions. With this approach he differed from most of his contemporaries.

If we reviewed other significant theologians of the latter part of the nineteenth century, such as ALBRECHT RITSCHL (1822 – 89) and WILHELM HERRMANN (1846 – 1922), or even ADOLF VON HARNACK (1851 – 1930), we would notice that the apologetic endeavor of Otto Zöckler was an exception. For most theologians the world hardly came into focus. God touches only the interior side of humanity. This retreat from the external and tangible was facilitated by the Kantian distinction between the phenomenal and the noumenal and also by Schleiermacher's claim that religion concerns itself primarily with feeling and intuition. Beyond that, the materialistic claim that everything existing can be reduced to matter and is subjected to an all-embracing cause-and-effect system left, in the eyes of most theologians, no other choice than to escape to something beyond the created order. With this retreat from the world, theology assumed increasingly a ghetto mentality.

3. British Empiricism and Its Consequences (17th–19th century)

In Great Britain two currents in the history of thought had a large following, on the one hand the *deism* of the Cambridge Platonists of which the most well-known representative was Ralph Cudworth (1617–1688) and on the other hand *empiricism* that was founded by John Locke. In his classic, *The True Intellectual System of the Universe* (1678) Cudworth attempted to refute any kind of atheism and materialism and propagated the unity of natural science, religion, and the Christian revelation as a medicine against atheism. This tendency was continued by Herbert of Cherbury (1583–1648) who also postulated the unity of faith and knowledge. In his book *De veritate* (*Concerning the Truth*; 1624) he advocated a natural religion which consists of basic characteristics common to all religions which also agree with reason. These five characteristics are:[1]

1. There is a highest deity.
2. We owe veneration to this deity.
3. Virtue coupled with piety shall be and has always been considered the main component of divine veneration.
4. Innate to the human soul is the abhorence of evil deeds. Therefore one is always cognizant that vices and crimes must be expiated through penance.
5. After this life there will be reward or punishment.

---

[1] Edward Lord Herbert of Cherbury, *De Veritate*, 3rd ed. [1645] ed. and intro. Günter Gawlick (Stuttgart-Bad Cannstatt, 1966), 210–220.

Herbert of Cherbury wanted to show that religion is based on a solid foundation of reason and therefore that it represents a reasonable monotheism. This was also the tendency of the so-called *free thinkers.*

The English philosopher ANTHONY COLLINS (1676 – 1729) pursued a similar intention as did Herbert of Cherbury. This becomes clear from his publication *A Discourse of Free-thinking* (1713). He claims that "ignorance is the foundation of atheism, and free-thinking the cure of it."[2] For him there are two sources from which religion flows, reason as the commonly acknowledged source and Scripture as the source of the specifically Christian way of expressing religion. His criticism of the Biblical texts was motivated by apologetics because he wanted to show that the Christian faith is not faith in any authority which would require sacrifice of the intellect. The Christian faith "is a vestment of natural religion which is perceived by rethinking but extremely effective in history and pedagogy and satisfying the inclinations of the masses for the mysterious."[3] This means the Christian faith in ethical and religious regard is nothing else but natural religion.

3.1 The Battle over Innate Ideas

Collins was a close friend of JOHN LOCKE (1632 – 1704) who often is considered the Father of English Empiricism, the second widespread intellectual movement of the 18th and 19th century. Locke rejects the notion of innate ideas, since all our ideas originate through external impressions and the reflection over them. Therefore *"the idea of God is not innate."*[4] Locke then continues: "It seems to me plainly to prove, that the truest and best notions

---

[2] Anthony Collins, *A Discourse of Free-thinking* [London 1713] with a German trans., ed. and intro. Günter Gawlick, with an essay by Julius Ebbinghaus (Stuttgart-Bad Cannstatt: Frommann, 1965), 105.

[3] So Emanuel Hirsch, *Geschichte der neuern evangelischen Theologie*, 1: 314, in his assessment of Collins.

[4] So John Locke in his *Essay Concerning Human Understanding* (1.3.7) collated and annotated Alexander Campbell Fraser (New York: Dover, 1959 [1690]), 1:95.

men have of God are not imprinted, but acquired by thought and meditation, and a right use of their faculties: since the wise and considerate men in the world, by a right and careful employment of their thoughts and reason, attained true notions in this as well as other matters."[5] Through the knowledge of reason that it is impossible for something reasonable to develop through pure accident, Locke deduces the necessity of God's existence.

The Scottish philosopher DAVID HUME (1711 – 1776) radicalized Locke's approach and attempted to purify empiricism from all non-empirical influences. For him sense experience was most decisive. In his *The Natural History of Religion* (1757) he concerned himself with religion and sought its origin in polytheism. "According to the natural progress of human thought, the ignorant multitude must first entertain some groveling and familiar notion of superior powers, before they stretch their conception to that perfect being, who bestowed order on the whole frame of nature."[6] This polytheism, however, is more superstition than knowledge about the relationship of things. The philosophers only arrive at a recognition of a spirit or a highest intelligence as the prime cause of everything. Contemplating creation, Hume affirms that "we must adopt, with the strongest conviction, the idea of some intelligent cause or author."[7] This indicates a teleological understanding of the world. As a strict empiricist he then asks what the religious principles have caused in the world and arrives at the condemning result: the moral principles may be good but the exertion is pernicious so that ultimately ignorance becomes the mother of piety. Already before that Hume had done away with metaphysics with the scathing judgment: "If we take in our hand any volume; of divinity or school metaphysics, for instance; let us ask, *Does it contain any abstract reasoning concerning quantity or number?* No. *Does it contain any experimental reasoning concerning matter of fact and existence?* No. Commit it then to the flames: For it can contain nothing but

---

[5] John Locke, *Essay Concerning Human Understanding* (1. 3.16), 1:105.

[6] David Hume, *The Natural History of Religion* (I), A. Wayne Colver, ed., in Hume, *The Natural History of Religion* and *Dialogues Concerning Religion* (Oxford: Clarendon Press, 1976), 27.

[7] Hume, *The Natural History of Religion* (XV), 92.

sophistry and illusion."[8] Yet it was too early for such a radical condemnation of religion and metaphysics and therefore in academic circles it was largely refuted. In 1761 all of Hume's writings were put on the Index by the Vatican. Nevertheless, people were curious to see what Hume had to say and bought his books so that he could live well from the sale of his publications.

Against this radical empiricism Kant wrote his *Critique of Pure Reason* (1781) in which he analyzed the possibilities of human reasoning "with reference to the cognitions to which it strives to attain without the aid of experience; in other words, the solution of the question regarding the possibility or impossibility of metaphysics."[9] Kant then arrives at a self-curtailment of reason because its actual interest cannot consist in regarding the ideas of freedom, immortality, soul, and God as objects of recognition but as objects of faith.

3.2 The Design Argument

In Great Britain there developed still another direction in which, following the lead of physico-theology, the attempt was made to prove God's existence from nature. It was especially WILLIAM PALEY (1743 – 1805), Archdeacon of Carlisle and a contemporary of Kant who pursued this notion.[10] Just like Leibniz he compared nature to a pre-established harmony which like a clock ran its course in pre-designed order. His *Natural Theology,* or *Evidences of the Existence and Attributes of the Deity Collected from the Appearances of Nature* (1802) enjoyed already in 1820 its 20th edition. His *Evidences of Christianity* (1794) appeared also in 1811 already in its 15th edition. As can be gathered from the success of his books many saw it necessary to combat empiricism

---

[8] David Hume, *An Enquiry Concerning Human Understanding* (12.3) ed. Tom L. Beauchamp (Oxford: University Press, 1999), 211.

[9] Immanuel Kant, *Critique of Pure Reason* (A XII), trans. J.M.D. Meiklejohn (EBook 2003).

[10] For Paley cf. D. L. LeMahieu, *The Mind of William Paley: A Philosopher and His Age* (Lincoln: University of Nebraska, 1976).

on its own field. Paley claimed that "the contrivances of nature surpass the contrivances of art, in the complexity, subtlety, and curiosity of the mechanism."[11] Paley wanted to prove that a designer must have created the world. He claimed that if a person would find a watch in an uninhabited desert, the person would have to conclude by the organization of the different parts of the watch that it had been created by a skilled watchmaker. The case is similar with nature. "Every manifestation of design, which existed in the watch, exists in the works of nature; with the difference, on the side of nature, of being greater and more, and that in degree which exceeds all computation."[12]

Two items are noteworthy: 1. Paley was fascinated with a mechanistic understanding of nature and therefore could compare the workings of nature with all the organs in a living being with that of a watch. He found that the proof for a purposeful organization in nature was so convincing that a materialistic interpretation of nature like that of Julien Offray de La Mettrie could be refuted. 2. Paley was sure that he had introduced a cumulative argument in the full sense of the term since each example he had adduced was sufficient by itself to refute any atheism. He claimed: "Were there no example in the world of contrivance except that of the *eye*, it would be alone sufficient to support a conclusion which we draw from it, as to the necessity of an intelligent Creator."[13] Each example taken by itself shows that there is a design in nature and that design necessitates a designer. Paley concluded "that amongst the invisible things of nature, there must be an intelligent mind, concerned in its production, order, and support. These points being assured to us by Natural Theology, we may now leave to Revelation the disclosure of many particulars."[14] Through natural theology we may conclude that there is a creator and sustainer while revelation then fills in the

---

[11] William Paley, *Natural Theology: or, Evidences of the Existence and Attributes of the Deity, Collected from the Appearances of Nature*, ill. James Paxton, add. notes V. John Ware (Boston: Gould und Lincoln, 1860), 13.

[12] Paley, *Natural Theology*, 13.

[13] Paley, *Natural Theology*, 44.

[14] Paley, *Natural Theology*, 295.

particulars. – As one can easily see, Paley's argument for a creator deduced from nature was not such an unassailable proof. With this argument a totally determined and mechanistic worldview proposed by de La Mettrie, which excluded an active God, could be defended. Yet Paley's *Natural Theology* served for many years as required reading in university studies. Its purpose was to show the students the purposefulness of the world and the foolishness of explaining it in mechanistic terms.

The conviction that God's working could be recognized from nature was widely held until the late 19th century in Great Britain, the skepticism of Hume notwithstanding, and often encouraged scientific research in nature. This becomes clear with the Baptist ANDREW FULLER (1774 – 1815), a popular author and pastor, who claimed about the Scriptures: "Though they give us no system of astronomy, yet they urge us to study the works of God, and to teach us to adore Him upon every discovery."[15] Some even attempted to reconcile development with God's creative activity, such as the Scottish Evangelical amateur geologist HUGH MILLER (1802 – 1856), who edited a journal of the *Free Church* and wrote: "God might as certainly have *originated* the species by a law of development, as he *maintains* it by a law of development; the existence of a First Great Cause is as perfectly compatible with the one scheme as with the other."[16] This means that various means were employed to come to terms with developmental theories as they became more and more popular in the 19th century and to connect them with the Christian faith.[17]

There was another facet, however, that made a rapprochement

---

[15] Andrew Fuller, *The Gospel Its Own Witness; or, The Holy Nature and Divine Harmony of the Christian Religion Contrasted with the Immorality and Absurdity of Deism,* in *The Complete Works of the Rev. Andrew Fuller,* ed. Andrew G. Fuller (London: Henry G. Bohn, 1848), 41.

[16] Hugh Miller, *Foot-prints of the Creator; or, The Asterolepsis of Stromness,* with memoir by Louis Agassiz (Edinburgh: Adam and Charles Black, 1861), 12.

[17] Cf. David N. Livingstone/D. G. Hart/Mark A. Noll, eds., *Evangelicals and Science in Historical Perspective* (New York: Oxford University Press, 1999), who show the various strategies employed by the evangelicals to come to terms with evolution.

between theology and science very difficult in Great Britain. Biblical scholarship, especially as it pertained to the discussion of creation versus evolution, was badly missing in Great Britain. In the German theological faculties, by contrast, especially at Göttingen, Berlin, and Halle, there existed a freedom of scholarship almost unknown in British institutions. The Old Testament could be studied critically and professionally without regard for the practical needs of the ministry. By the early decades of the nineteenth century, "the Old Testament was being studied in Germany with a critical intensity and professionalism matched only by that of the sciences in Britain and America."[18] For instance the philosopher and poet SAMUEL TAYLOR COLERIDGE (1772–1834) and the Hebrew scholar and co-founder of the Oxford Movement EDWARD B. PUSEY (1800–82) had both been introduced to Old Testament studies at Göttingen. In Great Britain many theologians were not adequately prepared for this new and critical approach to the Bible and therefore condemned such heresies outright.

One of the books that aroused a storm of controversy was *Essays and Reviews* (1860), written by theologians who employed higher criticism. "The book went through thirteen editions in five years, called forth in reply more than four hundred books, pamphlets, and articles, and caused Anglo-Catholics and Evangelicals at Oxford jointly to sponsor a declaration reaffirming the inspiration and authority of the Bible."[19] Under the leadership of E. B. Pusey, this so-called Oxford Declaration on Inspiration and Eternal Punishment was prepared on February 24, 1864, and sent to every clergy of the established church in England, Wales, and Ireland with the plea to sign it without delay. The authors of the *Essays* claimed to *"interpret the Scripture like any other book."*[20]

---

[18] So James R. Moore, "Geologists and Interpreters of Genesis in the Nineteenth Century," in *God & Nature: Historical Essays on the Encounter between Christianity and Science*, ed. David C. Lindberg/Ronald L. Numbers (Berkeley: University of California, 1986), 332.

[19] Moore, "Geologists," 341.

[20] For this and the following quote, see Benjamin Jowett, "On the In-

This meant that one has to ascertain first the literal use, namely, "the meaning which it had to the mind of the prophet or evangelist who first uttered or wrote, to the hearers or readers who first received it." This also meant for the authors that if one wanted to maintain the value of the Bible as a book of religious instruction, one should not try every possibility "to prove it scientifically exact at the expense of every sound principle of interpretation, and in defiance of common sense, but by the frank recognition of the erroneous use of nature which it contains."[21] While the Bible is still God's word, it is that in human form, and that human form is not beyond reproach.

With this approach the authority of the Bible was by no means abandoned, as shown in a sermon delivered at Oxford University by FREDERICK TEMPLE (1821–1902), the later archbishop of Canterbury and father of William Temple. He was one of the collaborators of this volume and stated in the sermon:

The student of science

> if he be a religious man, he believes that both books, the book of Nature and the book of Revelation, come alike from God, and that he has no more right to refuse to accept what he finds in the one than what he finds in the other. The two books are indeed on totally different subjects; the one may be called a treatise on physics and mathematics; the other, a treatise on theology and morals. But they are both by the same Author; and the difference in their importance is derived from the difference in their matter, not from any difference in their authority. Whenever, therefore, there is a collision between them, the dispute becomes simply a question of evidence.[22]

This kind of separation between the two books, of nature and of the Bible, while maintaining the common author, God, in the charged atmosphere that surrounded Darwin's *Origin of Species,*

terpretation of Scripture," in *Essays and Reviews,* ed. with an intro. Frederic H. Hedge (Boston: Walker, Wise, and Co., 1862), 416 and 417 respectively.

[21] Charles W. Goodwin, "The Mosaic Cosmogony," in *Essays and Reviews,* 238.

[22] Frederick Temple, "The Present Relations of Science to Religion" (1860), in *Essays and Reviews,* 494.

amounted for many to be a capitulation of theology. If nature no longer witnesses to design, then natural theology has lost its cause. This was dangerous, since in the *Origin of Species* (1859) Darwin advanced four major points:

1. There are random variations among species.
2. Populations increase at a geometrical rate, and as a result, there is a severe struggle for life at one time or another.
3. Since there are variations useful to organic beings, individuals with useful variations "will have the best chance of being preserved in the struggle for life."
4. Individuals with useful variations will pass on the beneficial traits to the next generation and "will tend to produce offspring similarly characterized."[23]

Darwin's *Origin of Species* was met with both eagerness and outrage. "The small first edition of 1,250 copies sold the day of publication, and a second edition of 3,000 copies soon afterwards."[24]

3.3 The Battle over Darwin

The watchword for controversy was "random variations," because with this concept the argument of an intentional design (by God) was effectively eliminated. Even the British geologist CHARLES LYELL (1797 – 1875) pleaded with Darwin to introduce just a little divine direction into his system of natural selection.[25] One either had to stay with religion, for which there was design, or with Darwin, who excluded design. SAMUEL WILBERFORCE (1805 – 73), bishop of Oxford, who also warned his clergy against the *Essays and Reviews,* launched a theological offensive against Darwin in an

---

[23] Charles Darwin, *On the Origin of Species,* in *The Works of Charles Darwin,* 15:92 [126/28].

[24] According to Charles Darwin, *Autobiography,* in *The Works of Charles Darwin,* 29:146 [122/24].

[25] According to William Irwin, *Apes, Angels, and Victorians: The Story of Darwin, Huxley, and Evolution* (New York: Time, 1963), 144 – 45.

article condemning Darwinism for contradicting the Bible.[26] On June 30, 1860, in an address at Oxford before the British Association for the Advancement of Science, Wilberforce postulated the exaggerated claim that Darwin had said "humanity is descended from monkeys," a claim Darwin never made.[27] To this charge the zoologist THOMAS H. HUXLEY (1825–95) retorted that he would rather be a descendant of a humble monkey than of a man who is misrepresenting those who search for truth. While the debate was played up more and more as time went on, this meeting only showed the basic difference in perception. If there was evolution and a basic cohesion among all living beings, the creation account could no longer be true. This is what the good bishop found reprehensible, and many others agreed with him.

For Darwin himself, reaching a conclusion regarding a certain worldview was not so easy to decide, since he had grown up with Paley's notion of design. The *Encyclopaedia Britannica* even states that Paley's *natural theology* "strongly influenced Darwin."[28] Yet when we check Darwin's own works, we find less than a handful of references to Paley. The two most enlightening ones are from his *Autobiography*, which show that even during Darwin's schooldays Paley's *Natural Theology* was universally accepted, and the young Darwin was no exception. Darwin writes:

> In order to pass the B.A. examination, it was, also, necessary to get up Paley's *Evidences of Christianity*, and his *Moral Philosophy*. This was done in a thorough manner, and I am convinced that I could have written out the whole *Evidences* with perfect correctness, but not of course, in the clear language of Paley. The logic of this book and as I may add of his *Natural Theology* gave me as much delight as did Euclid. The careful study of these works ... was the only part of the academic course which, as I then felt and as I still believe, was the least use to me in the education of my mind. I did not at any time

---

[26] Cf. for the following David C. Lindberg/Ronald L. Numbers, "Beyond War and Peace: A Reappraisal between Christianity and Science," in *American Church History: A Reader*, ed. Henry Warner Bowden/P. C. Kemeny (Nashville: Abingdon, 1998), 224.

[27] Cf. Irwin, *Apes, Angels, and Victorians*, 6.

[28] "Paley, William," *Encyclopaedia Britannica*, 2002.

> trouble myself about Paley's premises; and taking these on trust I was
> charmed and convinced of the long line of argumentation.[29]

Like every undergraduate student Darwin had been thoroughly
acquainted with Paley's writings. And even later in life he had still
high praise at least for the logic of the argument. Yet when he
started his career as a naturalist he arrived at quite different
conclusions. Now the lawfulness is no longer traced back to a
Creator but it is in nature itself. Darwin writes:

> The old argument of design in nature as given by Paley, which for-
> merly seemed to me so conclusive, fails, now that the law of Natural
> Selection has been discovered. ... There seems to be no more design
> in the variability of organic beings and in the action of Natural Se-
> lection, than in the course which the wind blows. Everything in na-
> ture is the result of fixed laws.[30]

Two items have remained controversial ever since Darwin wrote
these lines:

1. The rejection of design in nature.
2. That everything in nature is the result of fixed laws.

If there is no design in nature, and if everything is the result of
fixed laws, then the inevitable conclusion was that nature is self-
sufficient. There is no need of a creator or sustainer.

Darwin, however, wanted to be a pure scientist. He never really
endorsed any philosophical or metaphysical conclusions derived
from his scientific observations and theses. For instance, HER-
BERT SPENCER (1820–1903), an English philosopher and avid
follower and popularizer of Darwin's theory of evolution developed a cosmic theory of an all-embracing evolutionary process
and advanced his thesis of the "survival of the fittest."[31] In spite of
Spencer's affinity to Darwin, Darwin himself claimed to be

---

[29] Charles Darwin, *The Autobiography of Charles Darwin* [58/9], in *The
Works of Charles Darwin,* 29:101.

[30] Darwin, *Autobiography* [87/9], 29:120.

[31] Cf. Herbert Spencer, *First Principles* (New York: De Witt Revolving
Fund, 1958 [1862]), where he forcefully sets forth his notion of evolution.

simply a scientist who did not understand these philosophical inferences.[32] It was similar with Ernst Haeckel of whom Darwin, as we noted, was very impressed as a naturalist. But when Haeckel advanced his monistic theory embracing religion and science and declared the universe to be infinite and eternal, Darwin remained silent.

But Darwin did not exclude religion completely. He writes in his autobiography:

> Another source of conviction in the existence of God, connected with reason and not with the feelings, impresses me as having much more weight. This follows from extreme difficulty over the impossibility of conceiving this immense and wonderful universe, including man with his capacity of looking far backwards and far into futurity, as the result of blind chance or necessity. When thus reflecting I feel compelled to look to a First Cause having an intelligent mind in some degree analogous to that of man; and I deserve to be called a theist.[33]

Though Darwin could not see in the details of nature the "finger of God" Darwin did not completely abandon the notion of a creator in whatever form he conceived of such. Who doubts this should read the final pages of Darwin's *Origin of Species* where he expressly pointed to a creator.

---

[32] Cf. Charles Darwin in his letter to J. D. Hooker (June 23, 1863), in *The Correspondence of Charles Darwin*, Frederick Burkhardt *et al.*, eds., vol. 11 (1863), (Cambridge: Cambridge University, 1999), 504, where he writes about Spencer: "I could grasp nothing clearly. But I suppose it is all my stupidity; as so many think highly of this work."

[33] Darwin, *The Autobiography of Charles Darwin* [92 f.], in *The Works of Charles Darwin*, 29:123.

4. North America's Problem with Darwin

When one looks to North America most people think of the so-called Scopes' Monkey Trial at the beginning of the 20th century in which a public schoolteacher was found guilty and fined because he taught the theory of evolution. If one looks more closely, however then the discussion between theology and the natural sciences is much more multi-faceted. The U.S. diplomat, historian, and educator ANDREW D. WHITE (1832 – 1918) in his two-volume work, *A History of the Warfare of Science with Theology in Christendom* (1897), coined the metaphor of a war between theology and the natural sciences. He also mentions "the myriad attacks on the Darwinian theory by Protestants and Catholics."[1] A professor of American history at Columbia University, RICHARD HOFSTADTER (1916 – 1970), conveys the same sentiment when he writes: "The last citadels to be stormed were the churches."[2] FRANK HUGH FOSTER (1851 – 1935), professor at several American colleges and seminaries, in his posthumously published book *The Modern Movement in American Theology* (1939), comes much closer to the truth when he suggests: "In strict accordance with its own principles, the appearance of evolution on the theological stage and the perception of its importance for the philosophy of religion was a very gradual affair."[3] Indeed, initially

---

[1] Andrew D. White, *A History of the Warfare of Science with Theology in Christendom*, 2 vols. (New York: D. Appleton, 1897), 1:78.

[2] Richard Hofstadter, *Social Darwinism in American Thought*, rev. ed. (New York: George Braziller, 1969), 24.

[3] Frank Hugh Foster, *The Modern Movement in American Theology:*

theologians paid very little attention to science and were even
content to show that science would not hurt theology or they
considered theology as something quite different from science.
Only relatively late did a negative backlash against science arise,
especially against evolution. This was mainly fostered through
the contact with Continental theologians. When that happened
theologians seemed to withdraw more and more from the scientific world in order to hold on to their religious beliefs.

4.1 Euphoria in a Progressive World

At first theology also maintained the optimistic outlook which
largely prevailed in America. Scientific discoveries could not
threaten or even challenge the Christian faith. To the contrary, it
seemed that everything which could be discovered in nature
could only strengthen the faith. ARCHIBALD ALEXANDER (1772 –
1851), who taught as the first professor at Princeton Theological
Seminary, stated: "The Bible furnishes the full and satisfactory
commentary on the book of nature. With the Bible in our hands,
the heavens shine with redoubled luster. The universe, which to
the atheist is full of darkness and confusion, to the Christian is
resplendent with light and glory. The first sentence in the Bible
contains more to satisfy the inquisitive mind than all the volumes
of human speculation."[4] There is a difference between a Christian
and a non-Christian looking at nature. The Bible is not a literal
proof text for what we ought to find in nature, nor does the
universe necessarily lead to God as in natural theology. Similar to
what Johannes Kepler suggested, nature only underscores for the
Christian what he or she already knows about God; it declares the

---

*Sketches in the History of American Protestant Thought from the Civil War
to the World War* (New York: Fleming H. Revell, 1939), 38.

[4] Archibald Alexander, "The Bible: A Key to the Phenomena of the
Natural World" (1829), in *The Princeton Theology: 1812 – 1921; Scripture,
Science, and Theological Method from Archibald Alexander to Benjamin
Breckinridge Warfield,* ed. Mark A. Noll (Grand Rapids: Baker, 1983), 96.

beauty of God's creation. The Bible attests to that by saying that God indeed created the whole universe.

Even the creation account in Genesis 1 posed no problem for the Christian mind. Though general chronology had it that the earth existed only for a few thousand years, one could also interpret the word "day" to mean a geological period of indefinite duration. Therefore CHARLES HODGE (1797–1878) asserted confidently:

> As the Bible is of God, it is certain that there can be no conflict between the teachings of the Scriptures and the facts of science. It is not with facts, but with theories, believers have to contend. Many such theories have, from time to time, been presented, apparently or really inconsistent with the Bible. But these theories have either proved to be false, or to harmonize with the Word of God, properly interpreted. The Church has been forced more than once to alter her interpretation of the Bible to accommodate the discoveries of science. But this has been done without doing any violence to the Scriptures or in any degree impairing their authority.[5]

Hodge maintained that the Bible is God's word and therefore cannot conflict with the facts of science, since science deals with God's creation. Yet he conceded that scientific theories have been advanced that were inconsistent with the Bible and therefore had to be discarded sooner or later. He also admitted that the interpretation of the Bible had to be changed more than once to accommodate new scientific insights. For Hodge this did not undercut biblical authority, because whatever was factual on one side had to agree with what was factual on the other side, and vice versa. There was a unity of truth that could require a reinterpretation of Scripture. By and large this was also the reaction of the American public. There was nothing to be afraid of in science, since science meant progress, and that was certainly beneficial and could not be against God's will. Even Darwin's *Origin of Species* could not change this widespread sentiment. Moreover, evolution was not seen so much in biological terms,

---

[5] Charles Hodge, *Systematic Theology* (Grand Rapids: Eerdmans, 1952), 1:573.

but in social and economic terms, interpreted in the evolutionary framework of the British philosopher Herbert Spencer. In America Spencer was generally seen as a philosopher of progress and of continually further development.[6]

4.1.1 Evolutionary Theory in a Theistic Gown

In March 1860 ASA GRAY (1810 – 88),[7] who began his tenure as professor of natural history at Harvard in 1842, published a long and careful review of *The Origin of Species* in the *American Journal of Science and Arts*.[8] Gray, a member of the First Church in Cambridge (Congregational), freely admitted that not every-one would agree with Darwin's ideas. He mentioned, for instance, JAMES DWIGHT DANA (1813 – 95), the editor of the *American Journal of Science* and professor of natural history at Yale, who he was sure would not accept Darwin's doctrines (11). He also cited the outstanding Swiss-born American naturalist LOUIS AGASSIZ (1807 – 73), professor of zoology at Harvard beginning in 1848, who connected the phenomenon of origin and distribution of the species directly to the divine will and therefore was not able to accept Darwin's proposal of a "natural" origin and distribution of the species. Although Gray judged Agassiz "to be theistic to ex-cess," he suggested that there "need be no ground of difference here between Darwin and Agassiz" (14). Gray showed that tele-ologists such as Agassiz were quite selective. They conferred only particular facts to special design but left an overwhelming array of the widest facts as inexplicable. This meant that, taking the picture of nature as a whole into consideration, one could only say that it was so because it had so pleased the Creator to con-

---

[6] Brian Hebblethwaite, "Herbert Spencer", in *TRE* 31:650, writes that in America Spencer was "applauded as philosopher of capitalism".

[7] Cf. for the following Hans Schwarz, "The Significance of Evolutionary Thought for American Protestant Theology," *Zygon* 16 (1981): 262 – 79.

[8] Reprinted in Asa Gray, *Darwiniana: Essays and Reviews Pertaining to Darwinism* (New York: D. Appleton, 1876), 9 – 61. Page references to this work are placed in the text.

struct each plant and animal. Now Darwin proposed a theory that showed how each plant and animal was created, and therefore we could trust that "all was done wisely, in the largest sense designedly, and by an intelligent first cause" (53).

Gray admitted that Darwin's doctrine of "natural" selection could also be denounced as atheistic. Yet he cautioned that such statements should not be made on scientific grounds. Gray reminded us that Newtonian physics was already compatible with an atheistic universe. But he was convinced that "it is far easier to vindicate a theistic character for the derivative theory" (54). In conclusion Gray asserted again that Darwin's book is not a metaphysical treatise: "The work is a scientific one, rigidly restricted to its direct object; and by its science it must stand or fall" (56). Although the first edition of Darwin's book had left the matter unresolved, Gray suggested that Darwin probably had not intended to deny any creative intervention in nature. On the contrary, the idea of natural selection implied so many manifoldly repeated independent acts of creation that the whole process was considered "more mysterious than ever" (56). Before his review went to press, Gray saw the second edition of Darwin's book and noticed "with pleasure the insertion of an additional motto on the reverse of the title page, directly claiming the theistic view which we have vindicated for the doctrine" (61).

In his perceptive review two points gained special emphasis:

1. Darwin's theory of evolution was not a denial of religion but a scientific theory substantiated on scientific grounds and therefore to be refuted only on those grounds.
2. Darwin's theory did not diminish God's creative activity. If interpreted theistically, it even enhanced our understanding of the magnitude of divine creation.

In a series of articles that followed his review, Gray said Darwin's theory of descent, or any other such theory, should not yet be accepted as true and perhaps might never become truth. He insisted, however, that the same care should guide any non-acceptance of such a theory, that is, the claim that there are no secondary causes that account for the existence of the manifoldness of plants and animals. With these assertions Gray did not want to flee into aloof neutrality, but he wanted to make sure that scientific truth rests on unambiguous proofs. This stage he claimed had not yet been attained with an evolutionary theory. But he was certain that the theory of descent would become more and more probable, and if it were ever established, it would be done "on a solid theistic ground" (175).

Gray was convinced that natural science raised no formidable difficulties to Christian theism.[9] But we should not settle for a system of interpreting nature "which may be adjusted to theism, nor even one which finds its most reasonable interpretation in theism, but one which theism only can account for."[10] The latter, he assured, had been found in Darwinism. Of course, he conceded immediately that the opposite hypothesis is possible, namely, that there is no overall design in nature. Yet "the negative hypothesis gives no mental or ethical satisfaction whatever. Like the theory of immediate creation of forms, it explains nothing."[11]

Gray was evidently walking a tightrope here. He did not want to say that Darwin's theory offered a compelling belief in a personal divine being. But he also wanted to assure any possible doubters of Darwin's theory that a theistic interpretation was the only satisfying one. Since the ethical and mental satisfaction with this kind of interpretation evidently does not come from external (natural) evidence, or from authority of Scripture (supernatural evidence), it must rest with the individual. Thus Gray's theistic interpretation of Darwin's theory is a personal predilection though reinforced by the overwhelming consent of other scholars. However, it is not anchored in the necessity of nature or of the human individual (cf. Immanuel Kant) but rests on persuasion. Therefore it is a vulnerable argument if personal preferences change.

---

[9] Cf. Asa Gray, *Natural Science and Religion: Two Lectures Delivered to the Theological School of Yale College* (New York: Scribner's, 1880), 65.

[10] Gray, *Natural Science and Religion,* 91.

[11] Gray, *Natural Science and Religion,* 91.

4.1.2 John Fiske, the Interpreter of Spencer

JOHN FISKE (1842 – 1901), a popular lecturer, writer, and assistant librarian at Harvard (1872 – 79), was an ardent defender of evolutionary thought. As a junior at Harvard he already had the reputation of being a well-equipped Darwinian, and he was reprimanded by Harvard president CORNELIUS C. FELTON (1807 – 62) for reading the positivist philosopher AUGUSTE COMTE (1798 – 1857) in church.[12] Less than ten years later when Fiske was asked to give a series of lectures entitled "The Positive Philosophy" (1869 – 70) at the school — a change of presidency and educational goals having taken place in the meantime — it was clear that he had left Comte behind to adopt Spencer as his philosophical mentor. The lectures eventually evolved into a two-volume work, *Outlines of Cosmic Philosophy Based on the Doctrine of Evolution, with Criticisms on the Positive Philosophy* (1875). He even made a special trip to England to converse with Spencer, Darwin, Huxley, and others before publishing the work.

As with Gray's relationship to Darwin, Fiske was not a blind follower of Spencer. Spencer attempted to provide an interpretation of the cosmos from a purely scientific point of view, relegating all implications of evolution for understanding God to a secondary place. Fiske, however, wanted to show the religious side of the cosmic philosophy as well as the scientific one.[13] In arriving at a cosmic theism that left the anthropomorphic theism behind, he wedded theism much more closely with scientific data than Gray dared to. Fiske declared: "The existence of God—the supreme truth asserted alike by Christianity and by inferior historic religions — is asserted with the equal emphasis by that Cosmic Philosophy which seeks its data in science alone."[14] He gave this assurance: "Though science must destroy mythology, it

---

[12] For details cf. H. Burnell Pannill, *The Religious Faith of John Fiske* (Durham, N.C.: Duke University Press, 1957), 12.

[13] Cf. Pannill, *Religious Faith*, 22 – 23, esp. n. 68.

[14] John Fiske, *Outlines of Cosmic Philosophy Based on the Doctrine of Evolution, with Criticisms on the Positive Philosophy*, 2 vols. (Boston: James R. Osgood, 1875), 2:415.

can never destroy religion; and to the astronomer of the future, as well as to the Psalmist of old, the heavens will declare the glory of God."[15] Fiske's God, however, bears little resemblance to the God encountered in Psalms. This is illustrated by Fiske's statement: "There exists a POWER, to which no limit in time or space is conceivable, of which all phenomena, as presented in consciousness, are manifestations, but which we can know only through these manifestations."[16]

We are not surprised that such disembodied theism would not pass unchallenged by theologians.[17] But it is much more significant that both Spencer and Darwin, though pleased with Fiske's work, avoided any comments about the religious implications that Fiske had drawn. Darwin, for instance, told him, "I think that I understand nearly the whole — perhaps less clearly about Cosmic Theism and Causation than other parts," and then proceeded to emphasize that he, Darwin, was mainly an inductive and empirical thinker and therefore Spencer's deductions impressed him although they could not convince him.[18]

Like Gray, Fiske did not introduce to the American audience Spencer's philosophy or Darwin's theories but his own theistic interpretation of their work. This was, for instance, totally different from the materialistic interpretation of Darwin's theory by the agnostic Ernst Haeckel when he introduced it to his German audience. Fiske's deep concern and interest come to the fore especially well in a speech at the farewell dinner given to Spencer in New York on November 9, 1882, at the conclusion of his visit to the United States.

In the speech, entitled "Evolution and Religion," Fiske showed

---

[15] Fiske, *Outlines,* 2:416.

[16] Fiske, *Outlines,* 2:415.

[17] Cf. the extensive review article by B. P. Bowne, "The Cosmic Philosophy," *Methodist Quarterly Review* 58 (Oct. 1876): 678, where Bowne doubts "if the new doctrine will much advance the interest of either religion or science."

[18] Charles Darwin, "Letter to John Fiske, December 8, 1874," in *The Life and Letters of Charles Darwin, Including an Autobiographical Chapter,* ed. Francis Darwin (New York: D. Appleton, 1896), 2:371.

that Spencer's services to religion had been no less than those to science.[19] The reason for this was that the doctrine of evolution asserted "that there exists a Power to which no limit in time or space is conceivable, and that all the phenomena of the universe," material and spiritual alike, "are manifestations of this infinite and eternal Power."[20] This power, Fiske claimed, forms the basis of all religions. Yet the doctrine of evolution also has an ethical side. As Spencer had shown, moral beliefs and moral sentiments are products of evolution. Therefore, contrary to anybody today who would question the binding value of morals, Fiske affirmed: "When you say of a moral belief or a moral sentiment that it is a product of evolution, you imply that it is something which the universe through untold ages has been laboring to bring forth, and you ascribe to it a value proportionate to the enormous effort that it has cost to produce it."[21] Fiske shows that the theory of evolution has an intrinsic ethical dimension since right living is intimately connected with the whole doctrine of the development of life on earth. That is to say, what is right tends to enhance the fullness of life and what is wrong tends to diminish it.

4.1.3 Agassiz and Le Conte, a Cautious Reaction

Louis Agassiz was so deeply influenced by his teacher, the French naturalist Georges L. Cuvier, that he opposed the theory of evolution until his death in 1873.[22] While Agassiz admitted minor modifications within the species, he argued that the animals first called into existence were followed by a succession of creations

---

[19] John Fiske, *Excursions of an Evolutionist* (Boston: Houghton, Mifflin & Co., 1884), 294 – 305. For details of Spencer's enthusiastic reception in the United States, see *Herbert Spencer on the Americans and the Americans on Herbert Spencer,* comp. Edward L. Youmans (1883; reprint, New York: Arno Press, 1973).

[20] Fiske, *Excursions of an Evolutionist,* 301.

[21] Fiske, *Excursions of an Evolutionist,* 303.

[22] For details cf. the informed study by Hofstadter, *Social Darwinism in American Thought,* 17 f.

until the time "when, as the crowning act of the Creator, man was placed on the earth as the head of creation."[23]

It is significant that Agassiz did not oppose evolutionary theory on ideological grounds. He conceded that living beings could be the products or results of laws established by the Almighty or the work of the Creator directly.[24] But he insisted that one must decide on the basis of scientific facts between these two possibilities, the former held by the evolutionists and the latter he advocated himself. According to Agassiz, scientific investigation showed that there had been interruptions in the sequence of living species. The first set of animals had gone on multiplying up to a certain period or level "and then disappeared to make room for another set of animals, and so in their turn each set of newcomers had vanished to give place to others."[25] Since these successions did not occur by one generation making room for another but were promulgated by great disturbances in the natural course of events and extensive changes in the prevailing conditions of the earth, and since there was no indication that the animal world had grown from small and simple beings to its present diversity, Agassiz sided with catastrophism. Huge catastrophes wiped out whole animal populations, which then were replaced by new ones.

His rejection of Darwinism did not occur on theological or religious grounds. He was convinced that divine Providence was compatible with Darwinism. But he repudiated Darwin's theory for strictly scientific reasons. His student JOSEPH LE CONTE (1823–1901) attempted to update Agassiz by showing that Agassiz had actually laid the groundwork for the success of the evolutionary theory when he demonstrated the geological successions of different forms of animals and the embryonic recapitulation of these successions.[26]

---

[23] Louis Agassiz, *The Structure of Animal Life: Six Lectures* (New York: Charles Scribner, 1866), 6.

[24] Cf. Agassiz, *The Structure*, 91.

[25] Agassiz, *The Structure*, 91.

[26] Joseph Le Conte, *Evolution: Its Nature, Its Evidences, and Its Relation to Religious Thought*, 2nd ed. (New York: Appleton, 1892), 44. Surprisingly

While Agassiz was unwilling to accept Darwin's theory, Le Conte was less hesitant. Contrary to Fiske's flamboyant advocacy of evolutionism, Le Conte proceeded more cautiously. He distinguished between organic evolution and human evolution. The former, he taught, arises slowly according to the principle of natural selection (cf. 96 – 97). Since our spiritual nature would forbid a ruthless struggle for human survival, our only hope for human evolution would be in accord with the Lamarckian idea "that useful changes, determined by education in each generation, are to some extent inherited and accumulated in the race" (98).

For Le Conte the kingdom of God is not something soon to be attained in the evolutionary process, as Fiske made us believe. Evil, he said, will not soon be eliminated; but it "has its roots in the necessary law of evolution. It is a necessary condition of all progress, and pre-eminently so of moral progress" (373). Evil allows us a choice, and it makes us go forward to acquire virtue. When we hear, however, that "virtue is the *goal of humanity; virtue cannot be given, it must be self-acquired,*" we wonder whether these deliberations do not imply a similarly self-redemptive moralism as Fiske advocated (372). When we notice further that Le Conte understood God's sovereignty to work strictly within the limits of the laws of nature, we need not be surprised that initially theologians were rather hesitant to accept any evolutionary model of the world, fearing that it would endanger the truth of the Christian faith.[27] But within the learned community evolutionary ideas had become more and more acceptable. In the early 1860s the *Atlantic Monthly* published ex-

---

Le Conte also claimed that Agassiz rejected evolution since it conflicted with his religious convictions. That his rejection of Darwin's theory was based on religious grounds seems to be a misunderstanding, as I have shown above. The page numbers in the following text refer to Le Conte's *Evolution*.

[27] Joseph Le Conte, *Religion and Science: A Series of Sunday School Lectures on the Relation of Natural and Revealed Religion, or the Truths Revealed in Nature and Scripture* (London: Bickers & Son, 1874), 301, where he wrote: "God himself works in Nature only within the limits of law. He cannot do otherwise (I speak it with reverence), He cannot violate law, because law is the expression of his will, and his will is the law of reason."

positions by Gray on the Darwinian theory, and it also allowed Agassiz to present the opposite view. Much more on the side of Darwin and Spencer were *Appleton's Journal,* founded in 1867, and the successful *Popular Science Monthly,* started in 1872, which brought the evolutionary theory to more than ten thousand subscribers.[28] Soon college students' interest in English science (i. e., Spencer, Darwin, and Huxley) had replaced interest in English literature, and in 1872 an editorial in the *Atlantic Monthly* claimed that natural selection had "quite won the day in Germany and England, and very nearly won it in America."[29] But how did the religious community respond to the new evolutionary theories of Spencer and Darwin?

4.2 From Fear to Embrace: Protestant Theology and Evolution

A glance at the theological journals in America in the 1860s and 1870s reveals that there are very few categorical rejections of general evolutionary thinking. Nevertheless, many thought that the theory of natural selection was not well-founded. The main argument did not even come from theology, because theologians did not fight for the biblical truth against scientific insights. They rather formed their opinions by listening to noted scientists of their time and also by referring to their own knowledge of science. Often they admitted that the Darwinian Theory, even if it proven to be correct, would not be a threat to the Christian faith because it could interpreted theistically. This was also the opinion of Charles Hodge who nevertheless vehemently attacked Darwinism.

4.2.1 The Fears of Charles Hodge

Charles Hodge published a pamphlet in 1874 with the title *What Is Darwinism?* Its true title should have been: What is wrong with

---

[28] According to Hofstadter, *Social Darwinism,* 22.

[29] As quoted in Hofstadter, *Social Darwinism,* 23.

Darwinism? According to Hodge, Darwin's "grand conclusion is [that] 'man (body, soul and spirit) is descended from a hairy quadruped, furnished with a tail and pointed ears, probably arboreal in its habits, and an inhabitant of the Old World.'"[30] Yet Darwin did not say anything about the human soul, contrary to what Hodge implied. Darwin also would have rejected Hodge's suggestion: "In using the expression Natural Selection, Mr. Darwin intends to exclude design, or final causes" (41).

Though enjoying a certain degree of overkill in his argument, Hodge did not want to be unfair to Darwin. He conceded that Darwin explicitly and repeatedly admitted the existence of a creator. But then he chided him for not saying anything about the nature of the creator or of his relation to the world (27). With reference to complicated organs of plants and animals, Hodge asked:

> Why doesn't he say, they are the product of the divine intelligence? If God made them, it makes no difference, so far as the question of design is concerned, how He made them: whether at once or by a process of evolution. But instead of referring to the purpose of God, he laboriously endeavors to prove that they may be accounted for without any design or purpose whatever (58).

Like Agassiz, Hodge admitted that God could have made the living beings at once or gradually through the process of evolution. But unlike Agassiz, he did not fault Darwin for advocating evolution. What he rejected was the notion that evolution was explained in natural terms instead of supernatural ones. By explaining the evolutionary process in natural terms and by natural causes, Hodge implied that Darwin had effectively banished God from the world. It is important to note that Hodge distinguished here between "Darwinism," meaning the explanation of the development of the world without reference to God, and "evolution," the evolvement of the world through God's design (104). He realized that one could affirm evolution without admitting Darwinism.

The reason for Hodge's uneasiness with Darwinism is evident.

---

[30] Charles Hodge, *What Is Darwinism?* (New York: Scribner, Armstrong & Co., 1874), 39 f. Page numbers in the following text refer to this work.

"God, says Darwin, created the unintelligent living cell ... after that first step all else follows by natural law, without purpose and without design."[31] To remove design from nature is therefore the dethronement of God the creator. Thus Hodge reached this verdict: "The conclusion of the whole matter is, that the denial of design in nature is virtually the denial of God. Mr. Darwin's theory does deny all design in nature; therefore his theory is virtually atheistic; his theory, not he himself. He believes in a Creator."[32] Hodge's evaluation of Darwin culminated in the paradox: "A man, it seems, may believe in God, and yet teach atheism."[33]

If it really proved true, as Darwin had declared, that random variations were the cause of evolutionary change, then this had nothing to do with design. At most, God played dice with his creation. We should remember that even ALBERT EINSTEIN (1879 – 1955), who certainly was not theologically as conservative as Hodge, rejected spontaneity in nature and therefore had grave reservations about the Copenhagen interpretation of quantum mechanics. There must be a plan discernible in nature even for those who are not theists. If there were only random variations, then we could hardly speak of a creator who had designed the universe and everything that was within it.[34] Here ARCHIBALD ALEXANDER HODGE (1823 – 86), Charles Hodge's son and successor was wiser. Archibald no longer looked primarily at the random variations, but put more emphasis on the general course that evolution took and saw there "a providential unfolding of a general plan."[35] This meant for him that even the Darwinian

---

[31] Charles Hodge, *Systematic Theology,* 3 vols. (New York: Charles Scribner, 1871), 2:15.

[32] Hodge, *Systematic Theology* (1871), 2:173. Cf. Hodge, *What Is Darwinism?* 148.

[33] Hodge, *Systematic Theology* (1871), 2:19.

[34] Cf. David N. Livingstone, *Darwin's Forgotten Defenders: The Encounter between Evangelical Theology and Evolutionary Thought* (Grand Rapids: Eerdmans, 1987), 102, who states that while Hodge followed "the Scottish tradition in placing very definite limits on his adoption of natural theology, he remained convinced that the teleological argument was sufficient to establish the existence of God as an intelligent voluntary agent."

[35] So Livingstone, *Darwin's Forgotten Defenders,* 114.

Theory if one considered the overall concept, could be considered in theistic terms.

Yet it was not simply as defender of the argument of design that made Charles Hodge react so vehemently against Darwinism. When we see the authors he referred to in *What Is Darwinism?* then we get a further clue. He mentioned the British naturalist RUSSEL WALLACE (1823–1913), who together with Darwin proposed the theory of the origin of species by natural selection. Then he refers to Huxley, Büchner, Vogt, Haeckel, and Strauss. For instance, he quoted Haeckel as saying that Darwin's theory of evolution led inevitably to atheism and materialism.[36] Since Hodge was familiar with the Continental discussion about Darwin and the antireligious propaganda by people such as Vogt, Büchner, Haeckel, and Strauss, he was afraid that the same might happen in the United States. But his fears were unfounded for two reasons:

1. The evolutionary ideas that came to America were not so much those of Darwin as those of Spencer. Darwin never visited the United States as Spencer had done. (On his visit in 1882, Spencer was celebrated and treated like royalty.)
2. Neither Darwin's nor Spencer's theories were simply received in the United States without adaptation. As Hodge perceptively noted, Darwin's most fervent advocate in America, Asa Gray, though an avowed evolutionist, was not a Darwinian. He interpreted Darwin's theory theistically.[37] The same happened with Spencer's philosophy through the writings of Fiske. The United States was founded by people who had a religious vision, and the materialists and atheists there had no chance of turning evolutionary theory into an instrument that would advance their cause.

There was still another reason for the theistic reception of evolutionary thought in the United States. Most institutions of higher learning that would provide the platform for an intellectual exchange concerning evolution were church operated, or at least in some way affiliated with the church. In England and

---

[36] Hodge, *What Is Darwinism?* 95.
[37] Hodge, *What Is Darwinism?* 174–75.

especially on the Continent, however, they were mostly state owned and thus provided a more liberal intellectual environment unrestrained by ecclesiastical guidance.

In May 1874 Gray published an extensive review of *What Is Darwinism?* declaring that one should not blame a naturalist for leaving the problems of purpose and design to the philosopher and theologian.[38] Purpose on the whole, Gray asserted, was not denied but implied by Darwin. Gray was right when he surmised that Hodge's treatise "will not contribute much to the reconcilement of science and religion."[39] As a result of Hodge's pamphlet many people who had never read a line of Darwin became convinced that Darwin was the great enemy of the Christian faith. But by now the great opponent of evolution, Agassiz, had died (1873). In the 1874 edition of his *Manual of Geology,* JAMES D. DANA (1813 – 95), professor of natural history and geology at Yale and the leading figure among American geologists, endorsed the concept of natural selection, and GEORGE F. WRIGHT (1838 – 1921) of Andover helped Gray publish his *Darwiniana* (1876).

4.2.2 From Hesitancy to Enthusiasm (Dawson, McCosh, Beecher, Abbott)

That even conservatives had become amenable to evolution can be seen in the geologist J. WILLIAM DAWSON (1820 – 99), president of McGill University and president of both the American and the British Association for the Advancement of Science. He had once supported Hodge, and in 1890 stated in his book *Modern Ideas of Evolution as Related to Revelation and Science* that the current Darwinian and neo-Lamarckian forms of evolution "fall certainly short of what even the agnostic may desiderate as religion."[40] Yet he observed: "Creation was not an instantaneous process, but extended through periods of vast duration. In every

---

[38] *The Nation* (May 28, 1974), reprinted in Gray, *Darwiniana,* 266 – 82.

[39] Gray, *Darwiniana,* 279.

[40] J. William Dawson, *Modern Ideas of Evolution as Related to Revelation and Science* (New York: Fleming H. Revell, 1890), 226.

stage we may rest assured that God, like a wise builder, used every previous course as support for the next; that He built each succeeding story of the wonderful edifice on that previously prepared for it; and that His plan developed itself as His work proceeded."[41] Evolution was not objectionable as long as it was not Darwinian, that is, proceeding with blind force and blind chance, or Lamarckian, proceeding from the impact of the environment.

Even before Dawson, JAMES McCOSH (1811 – 94), a philosopher-theologian and president of Princeton College, had accepted evolutionary thought in Hodge's own backyard. McCosh was critical of Darwin's theory, especially of his attempt to attribute the whole evolutionary process to natural selection. He also doubted that humanity should be as closely associated with the animal kingdom as Darwin had claimed. But then he confessed:

> There are clear indications, in the geological ages, of the progression from the inanimate up to the animate and from the lower animate to the higher. The mind, ever impelled to seek for causes, asks how all this is produced. The answer, if an answer can be had, is to be given by science, and not by religion; which simply insists that we trace all things up to God, whether acting by immediate or by mediate agency.[42]

Here a leading figure of American Presbyterianism declared his acceptance of the Darwinian Theory. Yet he was not simply going with the times. As McCosh acknowledged, it had become known "that Darwin was a most careful observer, that there was great truth in the theory, and that there was nothing atheistic in it if properly understood."[43] But McCosh was also compelled by an evident pastoral concern:

> I have all along had a sensitive apprehension that the undiscriminating denunciation of evolution from so many pulpits, peri-

---

[41] Dawson, *Modern Ideas*, 230.

[42] James McCosh, *Christianity and Positivism: A Series of Lectures to the Times on Natural Theology and Apologetics* (New York: Robert Carter, 1871), 63.

[43] James McCosh, *The Religious Aspect of Evolution* (New York: Charles Scribner, 1890), vii.

odicals, and seminaries might drive some of our thoughtful young men to infidelity, as they clearly saw development everywhere in nature, and were at the same time told by their advisers that they could not believe in evolution and yet be Christians. I am gratified beyond measure to find that I am thanked by my pupils, some of whom have reached the highest position as naturalists, because in showing them evolution in the works of God, I showed them that this was not inconsistent with religion, and thus enabled them to follow science and yet retain their faith in the Bible.[44]

When the geologist and professor at Oberlin Theological Seminary in Oberlin, Ohio, GEORGE FREDERICK WRIGHT (1838–1921), who was also a friend of Asa Gray argued in summary review "*Recent Works Bearing on the Relation of Science to Religion. No. V: Some Analogies between Calvinism and Darwinism*" that Darwinism was a Calvinistic interpretation of nature since it was anti-sentimental, realistic, and to some extent fatalistic, this was a sign that evolutionary thought had become respectable.[45]

This became even more obvious when the most prominent preacher of that time, HENRY WARD BEECHER (1813–87), finally came out in favor of evolution. In *Evolution and Religion* Beecher declared that "the theory of evolution is the *working* theory of every department of physical science all over the world."[46] He claimed that it was taught in all schools of higher education and that children were learning it, since it was fundamental to astronomy, botany, and chemistry, to name just a few academic subjects. But Beecher insisted that evolution was "substantially held by men of profound Christian faith," and although theology would have to reconstruct its system, evolution would "take nothing away from the grounds of

---

[44] McCosh, *The Religious Aspect,* ix – x.

[45] George F. Wright, "Recent Works Bearing on the Relation of Science to Religion. No. V: Some Analogies between Calvinism and Darwinism," *Bibliotheca Sacra* 37 (1880): 76.

[46] Henry Ward Beecher, *Evolution and Religion* (New York: Fords, Howard & Hulbert, 1885), as reprinted in part in Gail Kennedy, ed., *Evolution and Religion: The Conflict between Science and Theology in Modern America* (Boston: D. C. Heath, 1967), 18.

true religion."[47] The reason for Beecher's confidence regarding evolution was his belief in two kinds of revelation: "God's thought in the evolution of matter" (nature) and "God's thought in the evolution of mind" (reason and religion).[48] Our task is to unite and to harmonize them, and then it will be noticed that the interpretation of evolution "will obliterate the distinction between natural and revealed religion, both of which are the testimony of God."[49] Beecher was convinced that there could be no disharmony between the God who was active in nature and the God disclosing himself in Scripture. But he even went one step further, a step that eventually caused protest from the conservative side, asserting that God disclosed himself as much in nature as in religion. Thus natural religion was revealed religion.

Under Beecher's influence LYMAN ABBOTT (1835–1922), Beecher's successor at *Plymouth Church* (Congregational) in Brooklyn, NY, joined the ranks of theistic evolutionists and contributed much through his sermons and journalistic efforts to the idea that Darwinism was acceptable to Protestant thought.[50] In his *Reminiscences* (1915) Abbott confessed that he studied Spencer in 1866 but not Darwin or Huxley, since he was not much interested in science.[51] In *The Theology of an Evolutionist* (1897), however, he called himself "a radical evolutionist" or "a theistic evolutionist."[52] We are immediately assured that he reverently and heartily accepts "the axiom of theology that a personal God is the foundation of all life" but that he also believes that "God has but one way of doing things; that His way may be described in one word as the way of growth, or development, or evolution, terms which are substantially synonymous."[53]

---

[47] Beecher, *Evolution and Religion,* 19.

[48] Beecher, *Evolution and Religion,* 15.

[49] Beecher, *Evolution and Religion,* 20.

[50] Ira V. Brown, in his interesting study, *Lyman Abbott, Christian Evolutionist: A Study in Religious Liberalism* (Cambridge: Harvard University Press, 1953), 141.

[51] Lyman Abbott, *Reminiscences* (Boston: Houghton Mifflin, 1915), 285.

[52] Lyman Abbott, *The Theology of an Evolutionist* (Boston: Houghton Mifflin, 1897), 9.

[53] Abbott, *Theology of an Evolutionist,* 9.

While Abbott noticed that all biologists were evolutionists, he also observed that not all were Darwinians, that is, not all regarded the struggle for existence and survival of the fittest as adequate statements of the process of evolution.[54] He understood evolution as the history of a process, and not an explanation adduced by giving causes. Therefore he accepted Fiske's aphorism: "Evolution is God's way of doing things."[55]

By the 1890s evolution had become a universal system and was also applied to the Bible. Here of course, the big problem was how to reconcile the story of the fall with the descent, or rather ascent, of humanity. Abbott discovered that, apart from Genesis 3, the story of the fall played no role in the Old Testament. Even in the New Testament there is no mention of it, except when Paul talks of the struggle between flesh and spirit. Abbott found that Paul's description of this struggle was effectively interpreted by "the evolutionary doctrine that man is gradually emerging from an animal nature into a spiritual manhood."[56] Abbott understood Paul to say that sin "enters every human life, and the individual 'falls' when the animal nature predominates over the spiritual."[57] Incarnation is then interpreted as the perfect dwelling of God in a perfect human being. For Abbott, Christ lived and suffered "not to relieve men from future torment, but to purify and perfect them in God's likeness by uniting them with God."[58] Christ did not appease God's wrath, he simply laid down his life in love that others might receive life. As was Beecher, Abbott was convinced that God, dwelling in the world, spoke through all its phenomena. Suddenly evolution not only had become acceptable to the Christian faith but also had become the tool with which the Christian faith and religion in general could be interpreted.[59]

---

[54] Abbott, *Theology of an Evolutionist,* 6 f., 19.

[55] Abbott, *Reminiscences,* 460, and many other places.

[56] Abbott, *Reminiscences,* 459.

[57] Abbott, *Theology of an Evolutionist,* 186.

[58] Abbott, *Theology of an Evolutionist,* 190.

[59] Cf. Washington Gladden, *Who Wrote the Bible? A Book for the People* (Boston: Houghton Mifflin, 1891), in which he attempted to demonstrate that the Bible had a "natural history" as well as a supernatural one.

4.2.3 Discerning the Mixed Blessing of Darwinism: Sumner and James

With relative ease Darwinism became accepted in America in a thoroughly theistic fashion. This was different from the bitter struggle over Darwin between the freethinkers and conservatives in Germany that carried well into the twentieth century. But actually it was not Darwin and his theory of natural selection that became accepted, but Spencer and his cosmic theory of an all-encompassing evolutionary process and of the survival of the fittest. For a young and expanding country like the United States, it was only fitting that the biological theory of Darwin became an appendix to the social, economic, and philosophical theory of Spencer.

The social Darwinism, or rather Spencerianism, of WILLIAM GRAHAM SUMNER (1840–1910), professor of political and social science at Yale, and of the industrialists JOHN D. ROCKEFELLER (1839–1937) and ANDREW CARNEGIE (1835–1919), is still active today when those on welfare are classified as lazy; or when, regardless of calls for hidden or overt government support, free enterprise is advocated as the best economic system; or when competition is believed to supply us indefinitely with oil and natural gas. According to its own principles, this kind of Darwinism will have to modify itself through pressure either from outside or from within; or if it does not change, it will be modified through the collapse of the socio-economic system. But this Darwinism, widely advocated by the so-called political conservatives, did not make much stir in theology. It has therefore been widely neglected by theologians since theology, which is usually exercised by members of the socio-economic establishment or the "fittest," benefits from it.

There is also a liberal Darwinism, which is perhaps even causally related to the first kind. This optimistic evolutionism considers development and evolution as God's way of doing things. As the philosopher WILLIAM JAMES (1842–1910) perceptively noted, "the idea of a universal evolution lends itself to a doctrine of general meliorism and progress which fits the religious needs of the healthy-minded so well."[60]

It is interesting that James, who first studied and then taught together with Fiske at Harvard, discovered the shortcomings of this new optimistic religion of nature, in which form Darwinism was introduced by Fiske, Beecher, and Abbott. James criticized it for its attempt to explain evil away instead of seeing it as an intrinsic part of existence. He correctly stated: "The method of averting one's attention from evil, and living simply in the light of the good is splendid as long as it will work."[61] And it did work as long as America was expanding and was still unaware of its boundaries and limitations. But with World War I and the Great Depression, things appeared in a different light.

Then many people discovered, as James did in 1902, that Christianity was not synonymous with the gospel of the essential goodness of humanity and of eternal Darwinian (better: Spencerian) progress. They remembered that Christianity was essentially a religion of deliverance, that we were called to die before we could be born again into real life.[62] People felt betrayed by the unjustified evolutionary optimism, and some demanded that evolutionary theories be outlawed altogether. The course of events might have been considerably different if evolutionary thought had not made its strongest impact on the American mind through Spencer and his interpreter Fiske, who declared that evolution was God's way of doing things. If it would have been through Darwin and his interpreter Gray, who confessed himself to be "a Darwinian, philosophically a convinced theist, and religiously an acceptor of the 'creed com-

---

[60] William James, *The Varieties of Religious Experience: A Study in Human Nature* (New York: Collier Books, 1961), 88. Cf. also Edward A. White's penetrating study, *Science and Religion in American Thought: The Impact of Naturalism* (Stanford, CA: Stanford University Press, 1952), 4 – 8, where White emphasizes the influence of William James and Reinhold Niebuhr in the rediscovery of the true significance of the Christian faith against optimistic evolutionism.

[61] James, *Varieties,* 140.

[62] Cf. James, *Varieties,* 141.

monly called the Nicene,' as the exponent of the Christian faith,"
both social Darwinism and the conservative backlash might
have been avoided.[63]

4.2.4 The End of the Gilded Age

We must remember how Darwin was received in America if we
want to assess properly the lasting impact of his ideas. Darwin's
evolutionary theory was introduced in America in a decidedly
theistic framework. This initially mitigated against the possible
clash with the tenets of the Christian faith concerning creation
and providence. The vast majority of American Protestant
theologians initially saw nothing in Darwin's theory that was
irreconcilable with the Christian faith, provided the theory was
scientifically acceptable and was clad in a theistic framework
that maintained a personal God who created and sustained the
world. In the wake of the expansion of the new American continent, Darwin's theory was seen as part of Spencer's comprehensive evolutionary theory, which also included socio-economic aspects. After its initial overwhelming success, this
idealistic and speculative system clashed with the reality of
radical evil and injustice exhibited in history and society.
Failing to distinguish between Spencer and Darwin, more
conservative theological minds began to react against the evolutionary theory in general; and some wanted to ban it from the
earth altogether.

The Social Gospel movement at the turn of the century still
accepted evolutionary categories in its attempt to address the
social injustices that accompanied the phenomenal expansion
of America by emphasizing the social dimension of sin. This is
evident in remarks by WALTER RAUSCHENBUSCH (1861 – 1918),
the most well-known representative of the Social Gospel
movement, who wrote: "Jesus was not a pessimist. Since God
was love, this world was to him fundamentally good. He realized
not only evil but the Kingdom of Evil; but he launched the

---

[63] Gray, Preface to *Darwiniana,* vi.

Kingdom of God against it, and staked his life on its triumph. His faith in God and in the Kingdom of God constituted him a religious optimist."[64] For Rauschenbusch, Jesus took his illustrations from organic life to express the idea of the gradual growth of the kingdom. He was shaking off catastrophic ideas and substituting developmental ideas.[65] The evolutionary, forward-reaching, and upward-moving process was central to the ideas of social betterment espoused by the Social Gospel. Yet Rauschenbusch also recognized that World War I "has deeply affected the religious assurance of our own time, and will lessen it still more when the excitement is over and the aftermath of innocent suffering becomes clear."[66] Although the progressive drive was deeply entrenched in the American spirit, there were ominous signs that affairs might not continue as usual. World War I had been a relatively short episode for America, since America entered it only at the tail end. But the many thousands of European immigrants pouring into America as a result of the war showed that the victory left many problems unsolved.

4.2.5 The Conservative Backlash and the Retreat of Theology

In America, conservative movements picked up significant momentum in the first decades of the twentieth century. For instance, the Temperance Movement of the nineteenth century, interrupted by the internal strife of the Civil War, gained amazing popularity and finally led to prohibition starting in January 1920. This was celebrated by evangelicals as a major victory against social evils such as poverty and the corruption of morals. A few years earlier the publication of a series of small volumes of essays entitled *The Fundamentals* (1910–15) had meant another breakthrough for the conservative cause. Against the ever-

---

[64] Walter Rauschenbusch, *A Theology for the Social Gospel* (1917; reprint, New York: Abingdon, 1945), 156.

[65] Rauschenbusch, *A Theology,* 220.

[66] Rauschenbusch, *A Theology,* 181.

growing influence of continental European theologians such as
ALBRECHT RITSCHL (1822–89), MARTIN RADE (1857–1949),
and ADOLF VON HARNACK (1851–1930), an influential group of
British, American, and Canadian writers presented the conservative stand in these essays. In this somewhat uneven series,
conservative but scholarly contributions were mingled with
dispensationalist articles. These contained extensive reference to
evolution and included one contribution with the characteristic
title "The Decadence of Darwinism." Publication of *The Fundamentals* was financed by two wealthy laypeople, and eventually
three million copies were distributed to pastors, evangelists,
missionaries, theology students, and active laypeople throughout
the English-speaking world. The five fundamentals testified to in
these volumes were the inerrancy of the Bible, the virgin birth, the
atonement, the resurrection, and the second coming of Christ.
While *The Fundamentals* could not stop the liberal trend by
rallying the conservative forces, it widened the gulf between the
two and spawned the name 'fundamentalists' for the conservative
side.

The fundamentalists' determination to stamp out, wherever
possible, teachings that appeared to contradict Scripture was
sooner or later prone to lead to a clash with the theory of
evolution. This clash was even more likely since not everyone
was preoccupied with progress. Large numbers of people
outside metropolitan centers and places of learning were virtually unaffected in their beliefs and habits by the intellectual
and cultural climate of the day. They lived in essentially the
same way, in the same world, and with the same beliefs as their
pioneer ancestors had. Their conservative mood needed only
to be rallied around a common cause, and they could form a
respectable force in society. One such rallying point proved to
be the teaching of evolution in public schools. Between 1920
and 1930 some thirty-seven anti-evolution bills were introduced in twenty state legislatures and passed in several
states such as Tennessee, Mississippi, and Arkansas. For instance, in Tennessee, fundamentalist groups had become
powerful enough to pressure the state legislature in 1925 to
adopt legislation making it unlawful to "teach any theory that

denies the story of divine creation of man as taught in the Bible."[67]

The anti-evolution issue came to a climax when, in the summer of the same year, high school teacher JOHN SCOPES (1900 – 70) of Dayton, Tennessee, was put on trial for violating the recently passed statute prohibiting the teaching of evolution in tax-supported schools. The trial gained lasting fame since two prominent personalities took sides in it. On the side of the law was WILLIAM JENNINGS BRYAN (1860 – 1925), three-time presidential hopeful and ardent champion of the fundamentalist cause; and on the side of the accused, CLARENCE DARROW (1857 – 1938), famous criminal lawyer and militant agnostic who ridiculed biblical literalism sharply. The trial aroused not merely national but international interest, and was accompanied by an immense amount of publicity. Although Scopes' conviction in the lower court was overturned by the Supreme Court of Tennessee on grounds that the fine had been improperly imposed, the effect of the publicity on the general public was to discredit fundamentalism. As time passed, fewer and fewer thoughtful people took seriously the categorical rejection of evolution by fundamentalists; and with only few exceptions this extreme stance has virtually disappeared from the American scene.[68]

Yet theology withdrew more and more from the dialogue with the natural sciences. The large majority of conservative and neo-orthodox theologians of the 20th century hardly had an interest to carefully dialogue with the claims of natural science with regard to evolution. When we briefly look at the most prominent representatives of neo-orthodoxy in America, REINHOLD NIEBUHR (1892 – 1971) and H. RICHARD NIEBUHR (1894 – 1962), we do not find any reference to evolution in their major writings. For instance, Reinhold, in his seminal work *The Nature and Destiny of*

---

[67] According to Clifton E. Olmstead, *History of Religion in the United States* (Englewood Cliffs, N.J.: Prentice-Hall, 1960), 549.

[68] There are occasional backlashes, however. In August 1999 a Kansas school board decided to eliminate evolution entirely from the curriculum. But these are clear exceptions. For the revival of creationism cf. the relevant pages in chapter 6.

Man (1941), makes no mention of evolution, Darwin, or Spencer. Referring to the modern view of humanity, he briefly describes the idea of progress as one which, after eliminating the Christian doctrine of sinfulness, relates "historical process as closely as possible to biological process and which fail to do justice either to the unique freedom of man or to the demonic misuse which he may make of that freedom."[69] Similarly in his essay "The Truth in Myths"(1937), he refers to the myth of creation and claims that one ought to distinguish between what is "primitive and what is permanent, what is pre-scientific and what is supra-scientific in great myths."[70] While he discerns the inadequacy of purely rational approaches to the world, he does not relate the scientific to the religious insights. He simply wants to keep each of them in check so that they do not conflict with each other.

H. Richard Niebuhr, in his widely read book *Radical Monotheism and Western Civilization* (1960), has a long chapter, "Radical Faith and Western Science," discerning a parallel structure between the closed-society faith in religion and the closed-society faith in science. He is not worried that science would conflict with the religious element in religion but rather with the dogmatic truth system of the closed-society faith. Niebuhr's argument could be interpreted to mean that belief in God the Creator and Sustainer of all things does not exclude the notion of evolution and indeed might even imply it. All this remains on the level of conjecture. He does not mention evolution or its main interpreters. Neo-orthodox theology was so intent on defining its own task of espousing God's Word that it neglected the actual dialogue with other disciplines. This approach, of course, is largely influenced by Karl Barth.

---

[69] Reinhold Niebuhr, *The Nature and Destiny of Man: A Christian Interpretation*, vol. 1: *Human Nature* (1941); new ed. (New York: Scribner, 1964), 24

[70] Reinhold Niebuhr, "The Truth in Myths," in *Evolution and Religion: The Conflict between Science and Religion in Modern America*, Gail Kennedy, ed., 93.

5. The Continental Fortress Mentality and the Gradual Turnaround (First Half of the 20th Century)

As has already been indicated, German Protestant theology especially had already in the 19th century largely withdrawn from a dialogue with the blossoming natural sciences. This tendency even increased until the middle of the 20th century. The voice of theology did not become silent but attempted to assert itself in contrast and in opposition to the knowledge of the world of the natural sciences in such a way that it emphasized something completely different from what was being said in the natural sciences. This resulted in two different views of the world. The natural sciences talked about nature and theology spoke about creation. This way theology attempted to show that the truth claim of its view could not be challenged by the natural sciences.

5.1 A Strict Demarcation (Karl Barth)

As co-founder and main representative of the so-called "dialectic theology", also known as neo-orthodox or neo-reformation theology, the Swiss theologian Karl Barth (1886 – 1968) rejected categorically any contact or dialogue with the natural sciences in his Doctrine of Creation. Within more than 2,700 pages he lays out his doctrine of creation in epic breadth. It is divided into four chapters whereby each chapter is already one hefty volume, *The Work of Creation*, *The Creature*, *The Creator and His Creature*, and finally *The Command of God the Creator*. While the last volume is easily detected to contain mainly ethics, one would at least assume that in the other three volumes there is some discussion with the natural sciences. Yet it is noticeable at once in the first volume that there has been a strictly theological narrowing down of the topic, because creation is understood from the perspective of covenant. In the preface Barth preempts any potential criticism of his procedure when he writes:

> It will perhaps be asked in criticism why I have not tackled the obvious scientific question posed in this context. It was my original belief that this would be necessary, but I later saw that there can be no scientific problems, objections or aids in relation to what Holy Scripture and the Christian Church understand by the divine work of creation. . . . There is free scope for natural science beyond what theology describes as the work of the Creator. And theology can and must move freely where science which really is science, and not secretly a pagan *Gnosis* or religion has its appointed limit.[1]

If we fear that with this dictum any discussion with the natural sciences is ultimately precluded, we may regain hope with the next sentence: "I am of the opinion, however, that future workers in the field of the Christian doctrine of creation will find many problems worth pondering in defining point and manner of this twofold boundary."

A theologian is only concerned with God and God's work, while a scientist researches the world. But both refer to the created order and interpret it either as world or as creation. Therefore one would assume that a debate should ensue between theology and science. Yet Barth wants nothing to do with a dialogue because, as he states later in the same volume, we are confronted with "a fundamental difference between the Christian doctrine of creation and every existent or conceivable world-view."[2] The Christian doctrine of creation cannot become a worldview, nor can it be supported by one, nor does it support one.

---

[1] Karl Barth, *Church Dogmatics,* vol. 3, *The Doctrine of Creation,* pt. I, ed. G. W. Bromiley/T. F. Torrance (Edinburgh: T. & T. Clark, 1958), ixf.

[2] Barth, *Church Dogmatics,* vol. 3, pt. I:343, for this quotation and the following.

> It cannot come to terms with these views, adopting an attitude of partial
> agreement or partial rejection. ... Its own consideration of these views
> is carried out in such a way that it presents its own recognition of its
> own object with its own basis and consistency, not claiming a better but
> a different type of knowledge which does not exclude the former but is
> developed in juxtaposition and antithesis to it.

Barth once more makes it clear that in the strict sense no dialogue
is possible with other worldviews, including the natural sciences.

Theology and natural science work and argue alongside each
other, they are phenomena which stand strictly parallel to one
another. Barth does not come to this conclusion out of ignorance
or laziness. In the second part of his *Doctrine of Creation* he deals
extensively with theological apologetics, especially with Otto
Zöckler, and shows that it attempted to defend the special place of
humanity in the world over against Darwin's theory of descent.
Barth has some good things to say about this attempt, but he
contends that we should not deceive ourselves, "that we have
attained to real man, to his uniqueness in creation."[3] One is left
with the phenomenon of humanity, but it does not reach its reality as is shown to us by God in Jesus.[4] To talk about creation
means for Barth to talk about God and God's activity, and to talk
about humanity means to talk about Jesus Christ, since in him we
encounter the creature chosen and elected by God.

Barth's strictly theological understanding of creation has
certainly been helpful and liberating for many because he has
shown theologians and lay people a space in which they can
unfold the theological doctrine of creation without being disturbed by scientific knowledge. Certainly this also helped scientists, who saw their faith in creation threatened by scientific
knowledge. The physicist GÜNTER HOWE (1908 – 1968) for instance emphasized that Barth's theology had been liberating for
him. At the same time, however, the doctrine of creation lost its

---

[3] Karl Barth, *Church Dogmatics,* vol. 3, *The Doctrine of Creation,* pt. II,
ed. G. W. Bromiley/T. F. Torrance (Edinburgh: T. & T. Clark, 1960), 94.

[4] Barth begins each section of his anthropology with some reflections on
Jesus; *i. e.*, he begins each section with a Christological foundation.

anchorage in the world, a world which is largely shaped by applied science (i. e., by technology). This led to a standstill in the dialogue. In a society shaped by the natural sciences, theology was increasingly considered to be on the fringe. Yet the dialogue with the natural sciences was not completely extinguished.

5.2 The Scottish Peculiarity (Thomas F. Torrance)

The Scottish theologian THOMAS F. TORRANCE (1913–2007) was decidedly influenced by Karl Barth but also showed a great interest in the natural sciences. Significant for him was not only the Reformed heritage of Karl Barth with whom he had studied, but similar to Hodge he was influenced by Scottish Realism which convinced him that external objects can be immediately perceived by intuition. Moreover, there were several scientists among his relatives against whom he competed methodologically. Therefore he acquired knowledge in the sciences, especially in physics and also in the philosophy of science.

According to Torrance theology was also a science. Therefore he emphasized almost in opposition to Karl Barth:

> There are not two ways of knowing, a scientific way and a theological way. Neither science nor theology is an esoteric way of knowledge. Indeed because there is only one basic way of knowing we cannot contrast science and theology, but only natural science and theological science, or social science and theological science. In each we have to do with a fundamental act of knowing, not essentially different from real knowing in any field of human experience. Science is the rigorous and disciplined extension of that basic way of knowing and as such applies to every area of human life and thought.[5]

"Theology is the unique science devoted to the knowledge of God, differing from other sciences by the uniqueness of its object which can be apprehended only on its own terms and from within the actual situation it has created in our existence in making itself

---

[5] Thomas F. Torrance, *God and Rationality* (New York: Oxford University Press, 1971), 91.

known."[6] Like all other sciences, theological science is a human inquiry. It is not quackery, but it will put all claims that purportedly are the results of theological thinking to the most severe test to make sure it does not miss its object matter, the knowledge of God.

Since the primary object of theological inquiry is the one God who is the source of all being and the ground of all truth, theology is concerned with wholeness and unity, which may set it apart from any other science.[7] Like other scientific disciplines, theological science refers to the externally given reality. Moreover, theology comes up in its investigations against a boundary beyond which it cannot penetrate and which it cannot pass without inconsistency and error. Even theological science has its limits. Yet in theology we have to do with a divine object that draws us to itself. From the very beginning our knowledge of God starts with a union and not a disjunction between subject and object. This does not mean that subjectivism reigns supreme in theology, but that we could never talk about God unless God first speaks to us. This also means for Torrance that a strictly natural theology is not possible. Contrary to Barth, Torrance is not afraid of using the term "natural theology." Yet he considers natural theology to be incomplete in and of itself and that it is only consistent if it is coordinated with positive theology, meaning that it has its proper place within "the embrace of the theology of God's self-revealing interaction with us in the world."[8] Its concepts and theorems lack meaning and cogency in and of themselves and only make sense when supplemented and interpreted from the level of divine revelation. The reason for his acceptance of natural theology lies in the fact that "theology by its very nature can be pursued only within the rational structures of space and time within which we are placed by God, through which he mediates to us knowledge of himself, and within which we may develop and articulate our

---

[6] Thomas F. Torrance, *Theological Science* (New York: Oxford University Press, 1969), 281.

[7] Cf. Torrance, *Theological Science*, 282.

[8] Thomas F. Torrance, *Reality and Scientific Theology* (Edinburgh: Scottish Academic Press, 1985), 64.

knowledge of him."[9] This means that theology is an enterprise within this world and must conduct its own business with continuing reference to the parameters which we also know from the other sciences.

It is no surprise that a doctrine of creation is of utmost importance for Torrance. Two items are significant here: God in God's transcendent freedom made the universe out of nothing; and in giving it a reality distinct from Godself but dependent on Godself, God endowed the universe with an inherent rationality making it determinate and knowable. Theologically this means that the world can be understood as existing because of God's creative work and that it is being maintained by God. Considered from the point of view of natural science, however, the world can be known and understood because of its immanent rationality and determinate character. Nature then can only speak ambiguously of God if taken by itself. Though "it may be interpreted as pointing intelligibly beyond itself to God, it does not permit any necessary inferences from its contingence to God."[10] There is no inference possible from the created to the creator. It is not possible to prove in a rational way that God exists and that God has created the world, since the rational connection between creation and God is grounded in God alone. A natural theology or a theology of creation that does not take its starting point with God and God's self-disclosure is doomed to failure.

The Christian doctrine of God who is the creator of an orderly universe, who brought it into existence out of nothing and continuously preserves it from falling back into chaos and nothingness, has significance far beyond the confines of theology. All empirical-theoretical inquiry rests upon this contingent character of the world. "Natural science assumes the contingence as well as the orderliness of the universe."[11] While the notion of contingence was already present in Greek thought, it was re-

[9] Torrance, *Reality and Scientific Theology*, 64.

[10] Thomas F. Torrance, *Space, Time, and Incarnation* (London: Oxford University Press, 1969), 59 f.

[11] Thomas F. Torrance, "God and the Contingent World," *Zygon* (December 1979), 14:329.

stricted there by the notion of the created as being the embodiment of divine reason as well as by the dualism between eternal form and accidental matter. This meant that for ancient natural science the empirical was regarded as something secondary. This is the reason it took so long for the contingent and empirical character to gain the upper hand in the development of science. It means that natural science with its modern emphasis on the factual and empirical is largely due to the Judeo-Christian understanding of creation.

Since modern natural sciences and the Judeo-Christian tradition are historically interdependent, Torrance can assert: "But since it is the contingence of the realities of the empirical universe upon God that gives them their intelligibility and enables us to grasp their natural and inherent structures, genuine interaction between theological science and natural science cannot but be helpful to both."[12] God is the creator of all things visible and invisible and the source of all rational order in the universe. Therefore both theological science and rational science operate within the same space-time continuum, and theological interpretation and explanation cannot properly take place without constant dialogue with natural science.[13] There are especially two points at which dialogue becomes necessary for theology: in its emphasis on incarnation and resurrection as its basis and on the all-embracing miracles upon which the Christian gospel rests, because at these points God had acted decisively within the natural order of things.

It becomes clear that Torrance does not want to relegate theology and science into two entirely separate realms. Natural science is concerned with the universe in its natural, contingent process, and theological science focuses on the acts of God which in creation brought those processes into being out of nothing and established them in their utter contingency. Both have to do with the created order, theology inquiring into its transcendent source and ground, and science researching the contingent nature and

---

[12] Torrance, "God," 346.

[13] Cf. Thomas F. Torrance, *Space, Time, and Resurrection* (Edinburgh: Handsel Press, 1976), 22 f., for the following.

pattern of that order.[14] There are especially three points which theology should take note of in its conversations with science.

1. There has occurred a basic change in the concept of reality. Here Torrance points quite often to the Copenhagen interpretation of quantum theory, saying: "The emergence of relativity theory has had to give way to a profounder and more differential view of reality in which energy and matter, intelligible structure and material content, exist in mutual interaction and interdetermination."[15]
2. We must take note of the relational concept of space and time and of
3. The multileveled structure of human knowledge. The various sciences can be regarded as constituting a hierarchical structure of levels of inquiry which are open to wider and more comprehensive systems of knowledge.

For Torrance this opens up the possibility of a fruitful dialogue without asserting an either-or mentality which excludes the assertions of either theology or science. Torrance is open to accepting the most recent insights of the natural sciences and can also engage in a fruitful dialogue with philosophers of science, such as MICHAEL POLANYI (1891–1976) and ALFRED NORTH WHITEHEAD (1861–1947).

5.3 A German Outsider (Karl Heim)

While Karl Barth and his followers always considered creation from a strictly theological viewpoint and never moved to the other side, there were a few theologians who dared to go beyond the boundary and to also conduct the dialogue in the territory of the natural sciences. One of the few German theologians who did this was KARL HEIM (1874–1958). He was deeply appreciated by the readers of his publications as well as by the students who crowded into his classes. When he took his first sabbatical enrolment at the

---

[14] Cf. Torrance, *Space, Time, and Resurrection*, 180.

[15] Torrance, *Space, Time, and Resurrection*, 185.

university dropped so much that he was asked never to take another sabbatical leave. His tremendous popularity notwithstanding, he was passed over with silence by university theology which was at that time dominated by Karl Barth. Beginning with his youthful work *Das Weltbild der Zukunft* (*The Worldview of the Future*; 1904) till the concluding volume of his six-volume work *Der evangelische Glaube und das Denken der Gegenwart* (*The Protestant Faith and Present Day Thinking*) Heim continually dialogued with the natural sciences. From the very beginning it was his intent "to trace back the events of the world to a primal date from where we can perceive the whole world."[16] Similarly to Karl Barth, this primal date for him was God who had disclosed God's self in Jesus Christ. Yet this did not lead to a separation between God and the world, because the supra-polar primal space of God is the ultimate condition for polarity which faces us in our fallen world and also the hope that this polarity will one day find its dissolution and completion in God.[17]

Starting with the reality of all being which is given through God all theological assumptions must be logically consistent with the basic assumptions of the natural sciences. This means they cannot contain or have as a conclusion the negation of scientific insights. But religious basic assumptions must also be logically independent of current scientific results and they cannot serve as conditions for results in the sciences.[18] Unless the natural sciences are turned into an ideology they cannot contradict the basic assumptions of theology. In volume 5 of *The Protestant Faith and the Present Day Thinking* with the subtitle *The Transformation of the Scientific Worldview* Heim shows that this "non-contradictability" is no theological presupposition but results from the transformation in the sciences as it occurred through Albert Einstein and Werner Heisenberg in the theory of relativity

---

[16] So Karl Heim, "Zur Einführung", in Heim, *Glaube und Leben: Gesammelte Aufsätze und Vorträge*, 2nd ed. (Berlin: Furche, 1928), 15.

[17] Cf. Ingemar Holmstrand, *Karl Heim on Philosophy, Science and the Transcendence of God* (Stockholm: Almquist & Wiksell, 1980), 134 f.

[18] Cf. Holmstrand, *Karl Heim on Philosophy*, 78.

and in quantum mechanics.[19] Karl Heim picks up the new understanding of space and time, the problem of determinism in modern physics, and the contemporary theories of evolution. It is peculiar that in his theological reflections he extensively considers the issue of miracles.

In the last volume, *The World: Its Creation and Consummation* (1952; Eng. trans. 1962) Heim deals first with the creation of the world and then its completion. First Heim expounds how the origin and end of the world is viewed from the perspective of science and then follows it with an interpretation from the perspective of the Christian faith. According to Heim theology begins exactly there where science and its methods stop,

> for science can only explain the world of today from an element which was already present at the beginning. The Bible has something to say about the question how this primeval element arose and its answer can briefly be summarized in one sentence which is also the title-heading of the first chapter of the Bible: "In the beginning God created the heavens and the earth."[20]

At the end of this volume Heim explained what moved him to write this six-volume series: "To proclaim the Gospel of the redeeming power of Christ to a world which to a large extent rejects and contests this Gospel."[21] He wanted to provide a universal and intelligible foundation for the gospel of Christ. The course of the world and the whole history of nature are transitory phenomena for him if viewed from the perspective of scientific knowledge. God will lift "the world out of its fallen state and restore it to its original supra-polar character."[22]

Karl Heim noted that many Roman Catholic apologists are engaged in the dialogue with the natural sciences. But "the main school of Protestant theological thought maintains an attitude of

---

[19] Cf. for the following Karl Heim, *The Transformation of the Scientific World View* (New York: Harper, 1953).

[20] Karl Heim, *The World: Its Creation and Consummation*, trans. Robert Smith (Philadelphia: Muhlenberg, 1962), 57.

[21] Heim, *The World*, 151.

[22] Heim, *The World*, 158.

critical aloofness and declares that the entire discussion is 'out-of-date'."[23] This remark is clearly directed against Karl Barth and his followers. Yet Heim asserts: "If Christianity is not to allow itself to be relegated to the ghetto, if it is convinced that it has a universal message for the entire world and that like Paul it is a 'debtor both to the wise and to the unwise' …, then there is not avoiding discussion between the upholders of the Christian faith and the students of the physical universe."[24] Until the second half of the 20th century Karl Heim, however, was a lone voice.

He did not just attempt to defend the Christian faith against the attacks of science. But he began his discourse with the basic assumption that there cannot be a contradiction between faith and knowledge since both are God's gifts to humanity. Therefore he could show how unnecessary these so-called points of conflict were and render them harmless. He also went one step beyond a mere dialogue asserting that a person who does not have faith in Christ but lives only by scientific knowledge must ultimately become a victim of temporality. Therefore he confronted his readers with the inescapable alternative: either Christ or nothingness, either Christ or despair. This alternative betrayed Heim's pietistic heritage to which he remained attached throughout his life.

5.4 A Roman Catholic Voice
(Pierre Teilhard de Chardin)

The work of the French Jesuit and paleontologist PIERRE TEIL-HARD DE CHARDIN (1881 – 1955) was discovered and readily accepted by Protestants in the 1950s. Perhaps the official Roman Catholic Church had also contributed to his late and posthumous fame. The Church had been deeply suspicious of Teilhard and his unconventional evolutionary thoughts. It was only due to the

---

[23] Karl Heim, *The Christian Faith and Natural Science*, trans. Neville Horton Smith (London: SCM, 1953), 5.

[24] Heim, *The Christian Faith and Natural Science,* 5.

forceful advocacy by his fellow Jesuits that he was finally re-
habilitated in the wake of the Second Vatican Council.

Teilhard is both a scientist and a theologian and priest. In his
self-understanding the latter gains the upper hand, because he
endeavors to find a new synthesis of his theological, physical,
paleontological, and paleo-anthropological findings with regard
to a Christian view of the universe and of humanity.[25] His theological starting point is the incarnation. God has entered this
world and will finally unite the world with himself. The universal
Christ found by Teilhard in the New Testament is "the organic
center of the entire universe."[26] He is the Alpha and Omega, the
beginning and end. If Christ is universal, Teilhard concludes,
then redemption and the fall "must extend to the whole universe"
and assume cosmic dimensions.[27]

Teilhard wants to overcome the old static dualism between
spirit and matter. He thinks this can be accomplished by viewing
spirit and matter together in a universe which is historically
advanced through an inward-guided evolution. There are four
stages of development: the *cosmosphere* as the origin of the
cosmos; the *biosphere* as the advancement of life; the *noosphere*
as "the Earth's thinking envelope,"[28] which is intensified through
a psychophysical convergence, an event Teilhard calls the plan-
etization; finally, the *Christosphere* emerges when the whole
cosmos is permeated by Christ and taken up in God. Important
for this is the omega point. It is, so to speak, the inner circle of the
universe which describes the end point of the development. God
incarnate is reflected in our noosphere. He is the reflection "of the
ultimate nucleus of totalization and consolidation that is biop-

---

[25] So very clearly Benz, *Evolution and Christian Hope: Man's Concept of
the Future, from the Early Fathers to Teilhard de Chardin,* 212, and cf. his
evaluation of Teilhard, esp. 224 ff.

[26] Pierre Teilhard de Chardin, "Note on the Universal Christ," in Teilhard, *Science and Christ,* trans. René Hague (New York: Harper & Row,
1965), 14.

[27] Teilhard, "Note," 16.

[28] Cf. Pierre Teilhard de Chardin, *The Heart of Matter,* trans. René Hague
(New York: Harcourt Brace Jovanovich, 1978), 31.

sychologically demanded by the evolution of a *reflective* living Mass."[29] As the divine converges with the material, the universe becomes personalized and the person (of Christ) becomes universalized by merging with all of humanity and all that is material.[30] The whole evolutionary activity is therewith centered in a process of union or communion with God.

For Teilhard religion and natural science are not opposites. "Even though the various stages of our interior life cannot be expressed strictly in terms of one another," they must nevertheless "agree in scale, in nature and tonality."[31] Science "uses certain exact 'parameters' to define for us the nature and requirements, in other words the physical stuff, of 'participated' being. It is these parameters that must in the future be respected by every concept of Creation, Incarnation, Redemption and Salvation." Teilhard does not want a Gnostic symbiosis between theology and science, but he is convinced that a convergence is coming, because theology and science internally hang together. He regards the scientific study of the world as an analytic pursuit that follows a direction which leads away from the divine reality.[32] At the same time, this scientific insight also shows a synthetic structure of the world and forces us to change our direction so that we turn back to the unique center of all things which is the Lord our God. Christians need not be afraid of the results of scientific research, whether in physics, biology, or history.[33] Often the analyses in the natural and historical sciences are correct. Yet they do not threaten the Christian faith. "Providence, the soul, divine life, are synthetic realities. Since their function is to 'unify,' they presuppose, outside and below them, a system of elements; but those elements do not constitute them; on the contrary, it is to those higher realities that the elements look for

---

[29] Teilhard, *The Heart of Matter,* 39.

[30] Cf. Teilhard, *The Heart of Matter,* 44 f.

[31] For this and the following quote, cf. Teilhard, *Science and Christ,* 221 ff., in a letter of November 2, 1947.

[32] So Pierre Teilhard de Chardin, "Science and Christ or Analysis and Synthesis," in *Science and Christ,* 21.

[33] For the following, including the quote, cf. Teilhard, *Science and Christ,* 35.

their 'animation.'" Science does not endanger faith, but helps us to know God better, to better understand and appreciate God. It does not make sense and it is unfair to pit science and Christ against each other or to separate the two into different realms which are foreign to each other.

On their own the natural sciences cannot discover Christ. The scientific endeavor gives birth to a yearning which Christ satisfies. Teilhard celebrates a cosmic Eucharist in a visionary or even mystical ecstatic way. In this process the divine fire permeates and illuminates the whole cosmos. However, Teilhard's idea of a Christification of the universe leads one to wonder if he has not used the traditional Roman Catholic understanding of transubstantiation as an unreflected premise.[34] It certainly may have influenced him. But much more important for Teilhard is the idea of evolution, which he understands in a totally Christocentric way. Similar to Karl Heim, Teilhard starts with the presupposition that there is no opposition between faith and thought, since both are gifts of God. Therefore, in an apologetic way Teilhard wants to issue a reasonable invitation to faith.[35] He does not start with physics, as does Heim, but with biology. It is his special contribution that he reaches out into that area about which the Roman Catholic Church has been suspicious for so long and attempts to penetrate it theologically.

5.5 A New Beginning of the Dialogue by the Natural Sciences

As seen above, theology had little interest in the dialogue with the natural sciences at least as far as Protestant continental theology was concerned. For scientists, however, the situation was very different. Leading scientists of that time like ALBERT EINSTEIN (1879–1955), WERNER HEISENBERG (1901–1976), or MAX

---

[34] Cf. Pierre Teilhard de Chardin, *The Divine Milieu: An Essay on the Interior Life* (New York: Harper Torchbook, 1965), 142 ff.

[35] Cf. also Henri de Lubac, *The Religion of Teilhard de Chardin,* trans. René Hague (London: Collins, 1967), 232.

PLANCK (1858 – 1947) were concerned with physics. In their work they were always confronted with boundaries which implied for them religious issues. This meant that it was not the theologians but the scientists who posed theological questions.

We can see this for example with GÜNTER HOWE (1908 – 68). After he had received his doctorate in mathematics he encountered Karl Barth whose theology opened for him new possibilities. He regarded Karl Barth's emphasis on the laity and their responsibility and also the starting point of theology with the Triune God as a quantum jump of history. Howe discovered analogous structures in Karl Barth's Doctrine of God and Niels Bohr's Theory of Complementarity.[36] Encouraged by the physicist and philosopher Carl Friedrich von Weizsäcker, Howe thought that a dialogue on these similarities would be fruitful. As a result of his enthusiasm approximately twenty-five theologians, physicists, mathematicians, and chemists came together in Göttingen in 1949 for a conference which then was repeated annually. These conferences in which philosophers were also later involved lasted until 1961. Then they were continued as an expanded enterprise in Heidelberg under the title *Evangelische Studiengemeinschaft* (*Protestant Community of Studies*).

Karl Barth excused himself for the first conference since he was too busy, and Karl Heim who would have loved to participate was at that time too sick to make the journey to Göttingen. Karl Barth regarded such a dialogue as a waste of time because in his opinion the old "natural theology" would have been moved again to the foreground. Therefore he would have only participated in a negative way and would have argued against the views of the scientists. Two of his former students who taught in Göttingen raised similar objections, ERNST WOLF (1902 – 1971) and OTTO

---

[36] Cf. Günter Howe, *Die Christenheit im Atomzeitalter. Vorträge und Studien* (Stuttgart: Ernst Klett, 1970), and esp. his contribution of 1958 "Das Göttinger Gespräch zwischen Physikern und Theologen" (The Göttingen Dialogue between Physicists), as well as the postscript by Hermann Timm, "Zu Günter Howes theologischem Lebensweg" (About Günter Howe's Theological Curriculum Vitae), and the preface by Carl Friedrich von Weizsäcker.

WEBER (1902–1966) so that FRIEDRICH GOGARTEN (1887–1967) became the Göttingen dialogue partner for the natural scientists. Because of his friendship with RUDOLF BULTMANN (1884–1976) it was not surprising that the conference in 1953 discussed Bultmann's thesis regarding the non-objectifiability of faith. Bultmann himself participated in that meeting.

In 1958 the Secretary General of the World Council of Churches WILLEM VISSER'T HOOFT (1900–1985) convened an international European conference in Bossey, Switzerland, in which the above-mentioned members of the Göttingen circle also participated. The development of atomic weapons had caused the direction of the dialogue to move toward the issues of a responsible science and a responsible society. Through his rising interest in a "responsible natural science" Günter Howe struck up a friendship with HEINZ-EDUARD TÖDT (1918–1991) of the University of Heidelberg. They staged joint lecture series and seminars and Howe was even called as an adjunct professor to the Theological Faculty in Heidelberg. He was a leader in the ecumenical discussions on peace. But then he died suddenly in 1968 at the age of 60.

Another scientist PASCUAL JORDAN (1902–1980) was Professor of Physics at the University of Hamburg. He ventured less directly into the area of theology. In a publication of 1963, *Der Naturwissenschaftler vor der religiösen Frage* (*The Scientist Confronts Religious Issues*) he showed in a simple but convincing manner "that all the barriers, all the walls, do no longer exist which an older natural science had erected along the way leading to religion."[37] Interest in such a dialogue did exist as is shown by the fact that this publication went through a second edition in less than a year.

That the dialogue gained momentum can perhaps be best seen with a physicist of the younger generation, A. M. KLAUS MÜLLER (1931–1995), who became known among theologians through his collaboration with WOLFHART PANNENBERG (*1928) in his *Erwägungen zu einer Theologie der Natur* (*Considerations Toward a Theology of Nature*). This publication resulted from the *Karlsruher Physiker-Theologen-Gesprächen der jüngeren Generation* (*Karls-*

---

[37] Pascual Jordan, *Der Naturwissenschaftler vor der religiösen Frage: Abbruch einer Mauer* (Oldenburg: Gerhard Stalling, 1963), 357.

ruhe Conversations of the Younger Generation Between Physicists and Theologians) a sequel to the Göttingen conferences which had also been initiated by Günther Howe. Müller contributed to the publication with an essay entitled "Über philosophischen Umgang mit exakter Forschung und seine Notwendigkeit" (*Philosophical Treatment of Hardcore Research and Its Necessity*). According to Müller a physicist "knows that he does physics, and whether and when he does a good job at it, but he is not sure what physics is."[38] To determine what physics is, he must compare it with something else. The philosopher who ponders about physics then looks for a horizon within which the truth of physics becomes visible. It follows that philosophical questions lead the one who deals with scientific research to get better into focus one's own horizon of consciousness. This means that new perspectives and new areas of research are opened up. Müller illustrates this using examples from the history of the sciences and emphasizes that a critical examination of their own presuppositions has often led to new progress.[39] Therefore a dialogue is necessary to clarify one's own thinking and to progress in it.

The next larger publication, *Die präparierte Zeit: Der Mensch in der Krise seiner eigenen Zielsetzungen* (*The Prepared Time: Humanity in the Crisis of Its Own Goals*) shows that humanity enters a dangerous time. It is interesting that Müller's book is dedicated to Günter Howe who along with Carl Friedrich von Weizsäcker was one of his tutors. One of his great challenges in writing this book was "the experience that the survival of humanity at the pinnacle of technological civilization is questionable yet even improbable."[40] Methodologically the principle of survival takes center stage in this book though Müller em-

---

[38] A. M. Klaus Müller, "Über philosophischen Umgang mit exakter Forschung und seine Notwendigkeit," in A. M. Klaus Müller/Wolfhart Pannenberg, *Erwägungen zu einer Theologie der Natur* (Gütersloh: Gerd Mohn, 1970), 9.

[39] Müller, "Über philosophischen Umgang," 16.

[40] A. M. Klaus Müller, *Die präparierte Zeit: Der Mensch in der Krise seiner eigenen Zielsetzungen*, intro. Wolf Häfele, pref. Helmut Gollwitzer (Stuttgart: Radius, 1971), 21.

phasizes a certain theory of knowledge which he sees embedded in the larger context of "biographical experience." This leads him to the theological concept of revelation for which he adduces Pannenberg's thesis that revelation occurs at the end of history and finally to eschatology for which again he relies on Pannenberg's thinking. He concludes his book with a comment that the Christian hope in a life beyond death "under the present conditions of survival seems to be the only hope which could set free a dynamic of life which could withstand the catastrophes of the coming decades."[41]

In a later publication, *Die Wende der Wahrnehmung* (*The Turning Point of Perception*) Müller summarizes the background for his earlier works in this way:

> To be in the world means to perceive. The body of perception is time. The nature of any perception depends on the nature of time. The nature of time is not timeless, but open to a very slow historical change. The nature of the time of the world which so far has been realized in the universe has formed the shape of knowledge: knowledge communicates in the difference of the modes of time in which time appears till today. Faith is a surrender to a being held above the abyss in which a future nature of time prepares itself.[42]

In *Die Wende der Wahrnehmung* Müller showed how our awareness of the world in which we live has changed in confrontation with the human ecological crisis. The first step to a new awareness consists in admitting the basic contradictions of our present understanding of the world instead of suppressing them. In his analysis of time and its changes one notices the influence of Karl Heim.

With Klaus Müller we have almost reached a representative of the present generation who is engaged in the dialogue. Though he treats classical issues he is concerned about our present situation. This means the dialogue is largely determined by the environ-

---

[41] Müller, *Die präparierte Zeit*, 646.

[42] A. M. Klaus Müller, *Wende der Wahrnehmung: Erwägungen zur Grundlagenkrise in Physik, Medizin, Pädagogik und Theologie* (Munich: Christian Kaiser, 1978), 9.

ment in which it takes place. This is also evident on the theological side. There it was the French Jesuit and paleontologist
Pierre Teilhard de Chardin who first focused the attention of
theologians on the natural sciences and who led to a cosmic
understanding of the world. We owe it especially to the Roman
Catholic Dortmund theologian KARL SCHMITZ-MOORMANN
(1928–1996) that Teilhard's works have been translated into
German. SIGURD DAECKE (*1932) investigated in his theological
dissertation *Teilhard de Chardin und die evangelische Theologie:
Die Weltlichkeit Gottes und die Weltlichkeit der Welt* (*Teilard de
Chardin and Protestant Theology: The Worldliness of God and the
Worldliness of the World*; Göttingen, 1967). He not only pointed
out the main advantages of Teilhard's theology but also investigated the relationship between God and the world as shown by
prominent Protestant theologians of the 20th century such as Paul
Tillich, Friedrich Gogarten, Dietrich Bonhoeffer, Gerhard Ebeling, Wolfhart Pannenberg, and Karl Heim. Daecke showed that
Heim and Teilhard have a common concern but in their work they
come up with opposite results. One year earlier JÜRGEN HÜBNER
(*1932) published his dissertation with the title *Theologie und
biologische Entwicklungslehre: Ein Beitrag zum Gespräch zwischen Theologie und Naturwissenschaft* (*Theology and Biological
Theory of Development: A Contribution to the Dialogue between
Theology and the Natural Sciences*; Munich, 1966) in which he
analyzed essential German theological contributions to the
theory of evolution since Darwin. Yet his main interest in this
publication is devoted to the personalism of Emil Brunner as a
possibility to proceed in the dialogue with biology. Thereby he
compares Brunner's method with that of Tillich and Heim. Later
Hübner showed much interest in Johannes Kepler and more recently in issues of the environment and bioethics.[43] He also de-

---

[43] Jürgen Hübner, *Die neue Verantwortung für das Leben: Ethik im
Zeitalter von Gentechnologie und Umweltkrise* (Munich: Christian Kaiser,
1986).

serves credit for documenting the newly commenced dialogue with the sciences in an extensive bibliography.[44]

In 1965 the dissertation of GÜNTER ALTNER (1936–2011) appeared on the topic *Schöpfungsglaube und Entwicklungslehre in der protestantischen Theologie zwischen Ernst Haeckel und Teilhard de Chardin* (*Faith in Creation and Theory of Development in Protestant Theology from Ernst Haeckel to Teilhard de Chardin; Zürich, 1965*). Altner offers a survey of the theological reactions to the Theory of Evolution. He is especially well-equipped for the dialogue since he received doctorates in both theology and biology. This double major seems to have sensitized him for present-day issues, because he published extensively on issues such as the peaceful use of nuclear energy and environmental problems. Though he objects to the use of nuclear energy he does not argue ideologically but carefully shows the still unsolved problems with nuclear energy. Concerning genetic manipulation his approach is also quite careful but with genetic manipulations of humans he pronounces an apodictic "No!"[45] In his book *Schöpfung am Abgrund* (*Creation Confronting an Abyss*) Altner unfolds the comprehensive context in which environmental issues must be treated. In so doing he also leads a dialogue with other scientists who are engaged in this area beginning with the American historian of science LYNN WHITE (1907–1987) and his German followers all the way to the process theologian JOHN B. COBB (*1925). According to Altner the Judeo-Christian tradition does not encourage a destruction of nature though the command to subdue the earth has often been misused. We are rather confronted with the consequences of a history of dis-

---

[44] Jürgen Hübner, *Der Dialog zwischen Theologie und Naturwissenschaft: Ein bibliographischer Bericht* (Munich: Christian Kaiser, 1987).

[45] Günter Altner/Günter Richter, eds., *Atomenergie – Herausforderungen an die Kirchen: Texte, Kommentare, Analysen* (Neukirchen-Vluyn: Neukirchener Verlag, 1977); Günter Altner, "Der Mensch als Geschöpf," in Konrad von Bonin, ed., *Menschenzüchtung – Ethische Diskussion über die Gentechnik* (Stuttgart: Kreuz-Verlag, 1985), 63.

obedience.[46] The question is whether theology and the church have the energy to change this course through which there emerges an ever greater gap between nature and history. Humanity has lost view of creation and must rediscover its responsibility for other creatures and discover new forms of cooperation between itself and nature. Technology and economy should not only serve human progress but also consider the well-being of the environment. Altner refuses a salvational egotism which only concerns itself with humans. Instead of pointing to the resurrection he calls for a theology of the cross which is characterized by denial and asceticism.[47]

---

[46] Günter Altner, *Schöpfung am Abgrund: Die Theologie vor der Umweltfrage* (Neukirchen-Vluyn: Neukirchener Verlag, 1974), 73.

[47] Altner, *Schöpfung am Abgrund*, 205.

6. A Vivid Dialogue with Many Voices

We notice that around the middle of the 20th century the dialogue between theology and the natural sciences grew with amazing speed. The existential issues which are addressed by Altner could not be solved by theologians or by scientists alone. Before we outline this dialogue and its major players we must make special mention of the work of one person who decidedly influenced the present dialogue, even causing it to flourish anew, Ian Barbour.

6.1 The Grand Senior of the Dialogue: Ian Barbour

IAN BARBOUR (1923 – 2013) received a Bachelor of Science degree from Swarthmore College (1943), a Master of Science degree in physics from Duke University (1946), and a PhD in physics from the University of Chicago (1950).[1] His theological studies were done at Yale University, where he received a BD in 1956. He started out teaching in the Department of Physics at Kalamazoo College in Michigan (1949 – 1953) then moved to Carleton College in Northfield, Minnesota, where he taught in the department of religion from 1955 to 1973 and became professor of religion and physics and the director of the program in science, ethics, and

---

[1] For details of his biography and the development of his oeuvre cf. the Heidelberg PhD thesis of Christian Berg, *Theologie im technologischen Zeitalter: Das Werk Ian Barbours als Beitrag zur Verhältnisbestimmung von Theologie zu Naturwissenschaft und Technik* (Stuttgart: W. Kohlhammer, 2002), 28 – 53.

public policy (1974–1989). At a time when it was not yet fashionable to engage in the dialogue between theology and the natural sciences in the 1960's he wrote *Issues in Science and Religion* which was also translated into Spanish (1971) and Chinese (1993). He soon realized, however, that most people were not interested in a dialogue that primarily dealt with issues concerning worldview. They were much more concerned with ethical issues arising from the applied sciences which confronted them in their daily lives. Therefore he wrote the sequel *Science and Secularity: The Ethics of Technology* (New York: Harper & Row, 1970). When he received the Templeton Prize in 1999 he gave part of the prize money to *CTNS* (*Center for Theology and the Natural Sciences*) in Berkeley, CA, to help finance the endowed chair for Robert Russell.

In his widely used textbook, *Issues in Science and Religion*, Barbour first traces the relationship between religion and science from the 17th century to the present. He focuses in the 17th century on Galileo and Newton and in the 18th century on the age of reason, at the romantic reaction, and the responses by Hume and Kant. In the 19th century Darwin occupies center stage, while in the 20th century Barbour moves from neo-orthodoxy to process philosophy. In a second part, Barbour lays out the different methods employed by science and religion, concluding that both science and religion are selective as to their focus of interest. Then he shows the differences in the methods of the two fields. For instance, the degree of personal involvement is greater in religion than in science, because "revelation in historical events has no parallel in science" and "the intersubjective testability of religious beliefs is severely limited as compared to that of scientific theories or even scientific paradigms."[2]

Barbour also maintains that the contrast between science and religion is *"not as absolute as most recent theologians and philosophers have maintained"* (268). Yet he does not want to merge religion and science into a new kind of natural theology, but endeavors to present a theology of nature. Therefore he focuses in a third part on physics and indeterminacy, on humanity and

---

[2] Ian Barbour, *Issues in Science and Religion* (Englewood Cliffs, N.J.: Prentice-Hall, 1966), 267. Page numbers from this work are in the text.

nature, on evolution and creation, and finally on God and nature, concluding with the process views of Whitehead and CHARLES HARTSHORNE (1897 – 2000). While Barbour shows an affinity to process theology in his own theology of nature, he "makes no attempt to set aside Christian beliefs" as he sees John Cobb doing in his Christian natural theology (453 – 54 n. 45). For Barbour "the context of theological discourse is always the worshipping community," and therefore "theology must start from historical revelation and personal experience." Yet it must also include a theology of nature that does not disparage or neglect the natural order (453). Since the scientific laws are selective, abstract, and often statistical, Barbour sees no problem stating that God acts in history and nature. But he does not see this activity occurring in such a way that God's contribution can be separated from other causes. To the contrary, God acts in, with, and under other causes. This position, according to Barbour, seems to have been most convincingly advanced by Whitehead's vision of God's persuasive activity at various levels in the processes of this world.

While secular humanity trusts science and not God to fulfill its needs, Barbour feels this attitude is shortsighted. Though our mastery and control of nature has "considerable biblical support," Barbour points out that "without the concomitant biblical ideas of care and respect for nature," that dominion can easily turn "into arrogance and ruthless subjugation."[3] Therefore, it is important to heed the biblical message that holds up "both an ideal of social justice and a model of man as a responsible self" lest our technological society become more and more unjust and also potentially self-destructive.[4] It is characteristic of Barbour that he considers both the dogmatic and the ethical aspect of the relationship between theology and science. He does not solely focus on how to relate God, creation, and humanity, but is also concerned with our ways of using science and of changing the world in which we live. Therefore it was not surprising that in his

---

[3] Ian Barbour, *Science and Secularity: The Ethics of Technology* (New York: Harper and Row, 1970), 7.

[4] Barbour, *Science and Secularity,* 140.

Gifford Lectures of 1989 – 91 Barbour spoke on both religion in an age of science, and ethics in an age of technology.

The first part of the Gifford Lectures (*Religion in an Age of Science*) resembles his earlier book *Issues in Science and Religion.* This time he offers no historical introduction, but starts immediately by comparing religion and science and also indicating four ways of relating science and religion: conflict, independence, dialogue, and integration. Then he moves to specific issues in the discourse between the two fields, such as physics and metaphysics, astronomy and creation, evolution and continuing creation, human nature, and God and nature. A special chapter is devoted to process thought because he still advocates the process model, believing that it "seems to have fewer weaknesses than the other models."[5] In the process model

> God is a creative participant in the cosmic community. God is like a teacher, leader, or parent. But God also provides the basic structures and the novel possibilities for all other members of the community. God alone is omniscient and everlasting, perfect in wisdom and love, and thus very different from all other participants. Such an understanding of God, I have suggested, expresses many features of the religious experience and the biblical record, especially the life of Christ and the motif of the cross. Process thought is consonant with an ecological and evolutionary understanding of nature as a dynamic and open system, characterized by emergent levels of organization, activity, and experience.[6]

It is characteristic of Barbour that he does not give a whole-hearted endorsement to process theology, but always argues carefully for the biblical basis of a theology of nature. Typical for process thought, however, Barbour rejects the notion of a creation out of nothingness, which means there was nothing prior to creation except God's own self. But much more important for him is that process thought avoids the dualisms of mind and body, of

---

[5] Ian Barbour, *Religion in an Age of Science: The Gifford Lectures, 1989 – 1991* (San Francisco: HarperSanFrancisco, 1990), 1:270.

[6] Barbour, *Religion,* 1:269.

humanity and nature, and of history and nature.[7] Also, the eco-
logical sensitivity of process theology is significant for Barbour.
This becomes especially evident in the second part of his Gifford
Lectures, which focuses on ethics.

The nineteenth century generally assumed, Barbour contends,
that science-based technology would automatically lead to progress and improvement in human life. Indeed, we have witnessed
that modern technology has increased food production, improved
health, provided higher living standards and better communication. But at the same time, environmental and human costs have
been increasingly evident. Therefore Barbour gives neither uni-
lateral endorsement nor a latent condemnation of technological
progress. As he considers critical technologies with regard to agriculture, energy, computers, and at the same time looks at human
and environmental values, it becomes clear for him that we must
redirect technology so that humanity and the environment are not
the victims but the beneficiaries of technological progress. Since
the churches have usually supported the status quo but have also
contributed to social change, Barbour is convinced that they "will
have to change drastically if they are to facilitate the transition to a
sustainable world."[8] More important for him, however, are certain
biblical images which still have significant power to evoke responses in humanity. There is first the prophet's commitment to
justice stemming from a belief in the fundamental equality of all
persons before God. Then there is the prophetic view that the
whole of creation is part of God's purpose, and that we are accountable for the way we treat all forms of life.

Barbour looks for a new vision of the good life. He takes his cues
from the biblical literature where the good life is identified with
personal existence in community and not with material pos-
sessions. The Bible recognized "the dangers of both poverty and
affluence" and advocated the dignity of the individual and "the
importance of interpersonal relationships." Here the Bible "offers a
distinctive view of *persons in community* which avoids both col-

---

[7] Cf. Barbour, *Religion*, 1:144 f., and 269.

[8] Ian Barbour, *Ethics in an Age of Technology: The Gifford Lectures,
1989–1991* (San Francisco: HarperSanFrancisco, 1993), 2:261.

lectivism and individualism" and also asserts the ideal of sim-
plicity.[9] Again, Barbour looks for process theology, which asserts
against a prevailing anthropocentrism "the interdependence of all
beings and the intrinsic value of nonhuman life."[10] While Barbour
sees that there are new opportunities for change toward a more just
and sustainable world, there are also still enormous obstacles, such
as individual and institutional greed, or the political power of
corporations and bureaucracies with a vested interest in the status
quo.

Barbour offers a balanced view both to the possibilities of dialogue between theology and science and also toward necessary
changes we must make in our perception of both theology and
technology. Barbour's lifelong engagement in the dialogue between
theology and science earned him not only academic recognition in
the English-speaking world, notably in North America, but also the
monetary award and concomitant public recognition of the Templeton Prize for progress in religion.

6.2 The Institutionalized Dialogue

While a dialogue is always carried on by individuals, they need a
place to meet and engage each other as well as the interested public.
Since the 1960s the public has noticed more and more the potential
dangers which could arise from scientific discoveries. In Europe, at
least in conjunction with the nuclear threat from the Soviet Union,
there arose a distrust in science and even a fear of technology which
to some extent continues to the present day. This climate resulted in
scientists and theologians coming together more frequently either
to generate joint publications or to organize conferences and panel
discussions. The churches in Germany have been especially exem-
plary in that respect because through their academies, i. e., conference centers, they furthered the institutionalization of the dialogue.
Some of these avenues for serious dialogue either in church institutions or through inaugurating associations we will now briefly

---

[9] Barbour, *Ethics,* 2:262.

[10] Barbour, *Ethics,* 2:263.

mention. They allowed lively encounters between theologians and scientists of quite different persuasions with regard to profession and worldview and thereby provided possibilities to discuss important questions.

6.2.1 Opportunities in Germany

In 1985 the *Gesellschaft für Evangelische Theologie* (*Society of Protestant Theology*) dedicated its annual meeting to the topic "Is Reconciliation with Nature Possible?" in which it addressed environmental issues.[11] On the occasion of the centenary of Charles Darwin's death CARDINAL KÖNIG of Vienna, Austria (1905 – 2004) convened a conference on the understanding of humanity in the context of evolution. Many well-known scientists, such as KONRAD LORENZ (1903 – 1989), Günter Altner, HOIMAR VON DITFURTH (1921 – 1989), and NIKLAS LUHMANN (1927 – 1998) presented papers at that conference.[12] A somewhat older publication again on the instigation of a theologian is the volume by the German theologian and historian ERNST BENZ (1907 – 1978), *Der Übermensch* (*The Superman*). Well-known scientists such as the Swiss anthropologist ADOLF PORTMANN (1897 – 1982), the founder of "scientific parapsychology" JOSEPH B. RHINE (1895 – 1980), and the German-Austrian pioneer of space exploration EUGEN SÄNGER (1905 – 1964) contributed as authors.[13] Another publication of 1984, more evangelical in its outlook, summarizes the results of three conferences of natural scientists who are close to the German Student Mission (*SMD*). The topic of that publication was Creation and Evolution.[14] Finally the *First European Conference on Science*

---

[11] Jürgen Moltmann, ed., *Versöhnung mit der Natur?* (Munich: Christian Kaiser, 1986).

[12] Rupert J. Riedl/Frank Kreuzer, eds., *Evolution und Menschenbild* (Hamburg: Hoffmann and Campe, 1983).

[13] Ernst Benz, ed., *Der Übermensch: Eine Diskussion* (Zürich: Rhein-Verlag 1961).

[14] Edith Gusche, Peter C. Hägele, and Hermann Hafner, eds., *Zur Diskussion um Schöpfung und Evolution: Gesichtspunkte und Materialien zum Gespräch*, 2nd ed. (Marburg/Lahn: SMD, 1984).

and Religion must be mentioned. It took place in the Protestant Academy at Loccum, Northern Germany, March 13–16, 1986 under the topic "The Argument About Evolution and Creation." Such prominent scientists and theologians as the biochemist and Nobel laureate MANFRED EIGEN (*1927), the British biochemist and theologian ARTHUR PEACOCKE (1924–2006), and the above-mentioned theologian Jürgen Hübner, participated in the conference. Since then this conference, which is held in English, has matured to a biennial institution and is known as the *European Society for the Study of Science and Theology (ESSSAT)*. It meets in various European countries and publishes its proceedings.

By mentioning *ESSSAT* we have already touched on organizations which have especially furthered the dialogue between theology and the natural sciences. *ESSSAT* seeks to respond to the need for dialogue between theology and science.[15] Its specific aims are to advance an open and critical communication between theology and science; to share the common knowledge; and to work on the solution of interdisciplinary problems. Membership is open to everyone interested in the dialogue regardless whether the person is an atheist or a conservative Christian. Yet the Society sees itself oriented on the Judeo-Christian tradition. Every second year it stages a conference in which approximately 200 scientists, philosophers, theologians, and historians from virtually all European countries, and even from overseas, participate. Places for these conferences included Rome (1992), Munich (1994), Edinburgh (2010), and for the 15th conference in 2014 Assisi.

In conjunction with these conferences the *ESSSAT* Research Prize of € 2000 will be awarded for an outstanding original contribution at the postgraduate or doctoral level. The 2014 Research Prize went to Patricia Bennett for her dissertation *Relationality and Health: A Transversal Neurotheological Account of the Pathways linking Social Connection, Immune Function, and Health Outcomes* at Oxford Brookes University, Great Britain. In 2008 the prize was received by Astrid Dinter (*1969), professor at

---

[15] For most of the following information cf. the homepage of *ESSSAT* http://www.esssat.org/index.php?option=com_content&task=view&-id=15&Itemid=1

Weingarten, Germany, for her second dissertation (*Habilitation*) which focused on computer as a catalyst in the identity formation of adolescents, and in 2006 Anne Runehov, Associate Professor of Philosophy of Religion, at Uppsala University, Sweden, received the prize for her dissertation on Brain research and mystical experience. Anne Runehov is also one of the editors of the *Encyclopedia of Sciences and Religion* (Dordrecht: Springer, 2013). Another prize awarded at these conferences is the *ESSSAT* Student Prize of € 1000 for an essay written in an academic context at graduate or undergraduate level. Then there is also a Communication Price for communicating the interactions of science and religion to the wider public.

ESSSAT publishes a newsletter, *ESSSAT News,* which appears approximately four times a year. It contains news from the Society as well as announcements of conferences, symposia, prizes, and events in other organizations involved in the dialogue. Then there is the book series *Issues in Science and Religion* published by T & T Clark in London. The first volume, *The Human Person in Science and Theology* appeared in 2000 and was edited by Niels Hendrik Gregerson, Willem B. Drees, and Ulf Görman, while volume 6, *How Do We Know? Understanding in Science and Theology* was published in 2010 and edited by Dirk Evers, Anje Jackelén, and Taede A. Smedes. Then there is also a yearbook, *Studies in Theology and Science,* which was started in 1994 and which dealt with complexity and self-organization. The most recent one, 2009-2010, deals with forms of knowledge, the relation between knowledge and religious belief, understanding in Science, and knowledge and meaning. *ESSSAT* maintains on its homepage also a very useful link to science and theology organizations in Europe and another one to organizations outside Europe.

Many of the scholars who engage in the dialogue between theology and science, and who are active in other organizations and functions, such as Antje Jackelén, Willem Drees, and Dirk Evers are active in *ESSSAT.* The current president of *ESSSAT* is the German-born Lutheran Archbishop of Sweden, ANTJE JACKELÉN (* 1955). Having received her doctorate at the University of Lund, Sweden, she became assistant Professor of Systematic Theology/

Religion and Science at the Lutheran School of Theology in Chicago in 2001, succeeding Philip Hefner. After her promotion to Associate Professor in 2003 she also became Director of the *Zygon Center for Religion and Science* until she was elected bishop of Lund in 2006 and finally inaugurated as archbishop of Sweden in 2014. She has continued to hold an adjunct professorship of theology and the natural sciences in Chicago, indicating her strong interest in the interface between theology and science. Her major publication, *Time and Eternity: The Concept of Time in Church, Science and Theology* (Philadelphia: Templeton: 2010; German: 2002), explores the possibilities of a relationally determined and eschatologically qualified concepts of time. To accomplish this she analyzes thousands of hymns which deal with time, surveys biblical and theological concepts of time, and delves into scientific theories. This updated 1999 version of her dissertation at Lund provides various aspects of a theology of time. Jackelén realizes that time is a relational and diverse phenomenon which includes structures and relationships, being and becoming, and a temporal openness which necessarily reaches beyond physics. This summary result shows the necessity of dialogue between theology and the sciences. But now we turn back to organizational structures so important for the dialogue.

The International Society for Science & Religion has no direct ties to Germany but rather to Great Britain. Since it is similar in its goals as *ESSSAT* we want to mention it here. The Society (*ISSR*) was established in 2002 for the purpose of the promotion of education through the support of inter-disciplinary learning and research in the fields of science and religion conducted where possible in an international and multi-faith context. Its first president was John Polkinghorne. The presidency subsequently passed to the South African cosmologist and mathematician George Ellis (*1939), to the British biologist Sir Brian Heap (*1935) and then to British historian of science John Hedley Brooke (*1944), and currently to the British psychologist and ordained clergy Fraser Watts. The central aim of *ISSR* is to facilitate a dialogue between the two academic disciplines of science and religion, one of the most important current areas of debate in terms of understanding the nature of humanity. While maintaining rigorous qualifications for membership (membership is through nomination by existing members only) the Society has now grown to over 140 members, including many of the leading scholars in the field of science and religion. The society incorporates, and counts among its members atheists as well as representatives from different faith traditions including Buddhism, Hinduism, Judaism, and Islam in addition to Christianity. Membership is also widely distributed geographically, with representatives from South Korea, India, Australia, New Zealand and South Africa as well as from Europe and America.

As a genuine German institution mention must be made of the *Forschungsstätte der Evangelischen Studiengemeinschaft Heidelberg* (*FEST*; *Research Institution of the Protestant Study Community at Heidelberg*). It devotes itself to special projects and is financed by the Protestant Church in Germany (*EKD*). It has a small, salaried staff of both scientists and theologians. Günter Altner, Jürgen Hübner, Jürgen Moltmann, and HORST W. BECK (*1933), a theologian and scientist, and many others have participated in projects of the *FEST*. The goal of this organization is to clarify the foundations of the sciences through an encounter with the Gospel. It also devotes itself to interdisciplinary research, issues of peace, of Marxism, and of legal and philosophical matters. Its various publications mainly serve the purpose of documenting its research.

The *Karl-Heim-Gesellschaft zur Förderung einer biblischchristlichen Orientierung in der wissenschaftlich-technischen Welt* (*Karl Heim Society to Further a Biblical Christian Orientation in a Scientific Technological World*) also deserves mention. It was founded in 1974 and has a circle of approximately 500 friends and supporters who receive twice a year the journal of the society, *Evangelium und Wissenschaft* (*Gospel and Science*). It contains essays, and information about conferences and book reviews. Under its president, the theologian HANS SCHWARZ (*1939) the Society continues the legacy of Karl Heim and maintains the Karl Heim archives in the Bengel Student Dormitory (*Bengelhaus*) in Tübingen. Through seminars and annual meetings the Karl Heim Society furthers the dialogue between theology and the natural sciences. Financially it is supported by members, friends and not the least by the Protestant Church in Württemberg. Since 1989 it has published a yearbook, *Glaube und Denken* (*Faith and Thought*) in which scientists and theologians from various countries advance the dialogue between theology and the natural sciences. Every two years it awards the Karl Heim Prize for excellent scientific work in the interdisciplinary dialogue. In 2010 Rebekka Klein from the University of Heidelberg received the prize for her dissertation on sociality as a human condition. The prize in 2012 went to Andreas Losch for his dissertation who had already received an *ESSSAT* Student Prize in 2006. This shows that societies which engage in the dialogue often apply similar criteria for excellence.

In 1981 Horst W. Beck left the Karl Heim Society to assemble his own circle of friends and supporters. Through seminars and conferences he attempted to create a conservative alternative to the Karl Heim Society. He published *Wort und Wissen. Impulse, Materialien, Kurse für christliche Alternativen in Wissenschaft, Technik, Gesellschaft* (*Word and Knowledge: Impulses, Materials, Courses for Christian Alternatives in Science, Technology, and Society*). This venture is supported by the *Studiengemeinschaft Wort und Wissen* (*Study Community Word and Knowledge*) in Baiersbronn, Württemberg. The main slant of this study community is anti-evolutionary and pro-creationistic. This society is an alliance of Christians predominantly with jobs in science who advocate a biblical doctrine of creation. The recognition of human beings as creatures of God, and of the whole cosmos as God's creation pertains, according to their conviction, to all the sciences which deal with humanity and the whole creation. They are largely oriented toward creationism, an interpretation of creation based on a literal reading of the first few pages of the Bible. In this regard they also accept the insights of the intelligent design tradition.[16] – In the area of economics they deal finally with the view of humanity and ethics in economy. For instance they published a textbook by REINHARD JUNKER (*1956), a the-

---

[16] Henrik Ullrich and Reinhard Junker, eds., *Schöpfung und Wissenschaft: Die Studiengemeinschaft Wort und Wissen stellt sich vor* (Dillenberg: Hänssler-Verlag & CV – Verlag, 2008).

ologian and natural scientist, and SIEGFRIED SCHERER (*1955), a microbiologist, which is critical of evolution.[17] Evidence that such publications are well received by a wider reading audience is shown in that the book is already in its sixth edition.

Horst Beck and his friends do not avoid contact with those who think differently. This is exemplified by a conference in 1984 in the Protestant Academy at Hofgeismar where Horst Beck, Jürgen Hübner and GÜNTER EWALD (*1929), a mathematician who closely cooperates with the Karl Heim Society and the *FEST*, presented papers. In contrast to the situation in the USA the German *Volkskirche* (*people's church*) allows the coming together of people with quite different views to engage in dialogue between theology and the sciences, rather than becoming a comfortable sectarian-like ghetto of like-minded persons.

This diversity also shows in the *Institut Technik-Theologie-Naturwissenschaften* (*Institute Technology-Theology-Natural Sciences; TTN*) at the Ludwig Maximilian University in Munich. Through the efforts of the Protestant ethicist TRUTZ RENDTORFF (*1931), representatives from the Protestant Church, the natural sciences and industry, gathered to establish the *TTN* in 1992 with the intention of furthering the dialogue between technology, theology, and the natural sciences. From a theological standpoint the institute deals with ethical issues caused by the development of new technologies and of scientific research and the ensuing societal challenges for industry. Especially young scientists from different fields can conduct interdisciplinary research there and establish contacts with mentors and interested persons. The institute understands itself as a platform where interested persons from different scientific fields and contexts of practical application can enter into a trustworthy and open dialogue about ethically problematic issues. The institute is associated with the university but is financed through a subsidy from the Bavarian Lutheran Church, annual dues from its members, and financial contributions from industry. The latter source, however, does not yield as much financial support as one might expect, so the in-

---

[17] Reinhard Junker and Siegfried Scherer, *Evolution – Ein kritisches Lehrbuch*, 6th updated and expanded edition (Gießen: Weyel, 2006).

stitute depends largely on grants from foundations. Nevertheless, the institute has an important function in dealing with ethically sensitive issues especially by evaluating potential problems caused by applied technology. While it is still difficult in Germany to acquire private funds to foster the dialogue between theology and the natural sciences this is widely practiced in the USA. Therefore we turn now to that country.

Yet before doing so we must mention one more platform for the dialogue between the natural sciences, philosophy, and religion, the *Religion and Science Network Germany (RSNG)*. The *Religion and Science Network Germany* understands itself as a platform for the dialogue between the natural sciences, philosophy, and religion for interested groups and individual persons. Since 2005 it stages an annual congress in Hohenheim where the Academy of the Roman Catholic Diocese Rottenburg-Stuttgart maintains a conference center which supports this venture. At these annual congresses both established scientists and young scholars from different disciplines and persuasions gather to listen to presentations, exchange ideas and are informed about the possibility of financial subsidies. These annual conferences allow for co-operation between scientists, research groups, and publishers but also generate ideas for future conferences and research projects. *RSNG* is closely associated with the *Metanexus Institute* in the USA under its founder and director William Grassie (*1957). This institute wants to advance a global network initiative and it has received sizeable grants from the Templeton Foundation of which we will hear later. Main topics for the annual congress of the *RSNG* have included the challenge of naturalism, understanding humanity through the history of nature, neurosciences in the interdisciplinary dialogue, evolutionary biology, methods and range of the natural sciences, and others. *RSNG* also maintains a valuable list of the various groups in Germany who deal in the widest sense with issues in science and religion (theology) and of conferences and talks on these issues. But now let us turn to the dialogue in the USA.

6.2.2 "America, You Are Better Off" (Goethe)

In America also the dialogue was not begun by institutions but by the initiative of individual people such as ROBERT JOHN RUSSELL (*1946). On the campus of the Graduate Theological Union in Berkley, California in 1981 Russell, a theologian and scientist, founded the *CTNS* (*Center for Theology and the Natural Sciences*) with the intention of advancing the interaction between theology and the natural sciences. Its main focuses were research, instruction, and public awareness.

Through his involvement Russell won the support of the *Templeton Foundation* for many ventures of *CTNS*. This gave *CTNS* an immense public exposure. The *Templeton Foundation* is named after SIR JOHN MARKS TEMPLETON, SR. (1912 – 2008) who founded an investment fund in 1954 and after its tremendous financial success he started the *John Templeton Foundation* in 1987. Presently this foundation is continued by his son JOHN MARKS TEMPLETON, JR. (*1940).[18] Besides other philanthropic ventures the *Templeton Foundation* has awarded annually since 1972 the so-called Templeton Prize in the amount of one million English pounds. In the last few years the prize has gone mainly to natural scientists who concern themselves with the scientific angle of religious or theological issues. An example is JOHN D. BARROW (*1952) in 2006 who, together with FRANK TIPLER (*1947), published a sizeable investigation of the anthropic principle.[19] Through such financial incentives the dialogue between theology and the sciences receives an incentive the strength of which should not be underestimated.

The *Templeton Foundation* also finances three large international programs which are administered by *CTNS*. Robert Russell reported that from 1998 to 2002 there was a "Science and Religion Course Program" in which 326 grants of 10,000 US dollars each

---

[18] For further information see Robert L. Hermann, *Sir John Templeton: Supporting Scientific Research for Spiritual Discoveries*, 2nd exp. ed. (Philadelphia: Templeton Foundation Press, 2004), 160 – 166.

[19] John D. Barrow/Frank Tipler, *The Cosmological Anthropic Principle*, pref. John A. Wheeler (Oxford: University Press, 1988).

were given to individuals to start new courses in science and theology at universities, colleges, and theological seminaries throughout the world.[20] In addition 178 conferences, workshops, and town discussions were held in which 2,670 lectures were presented which reached an audience of approximately 30,000. More than 80 of these lectures were published. From 1996 to 2003 there was a program under the title *Science and the Spiritual Quest* through which 120 international scientists were invited to conferences to fathom the connection between the spiritual and the scientific dimensions. In nine countries events were staged which reached approximately 400 million people through the media. Finally, in 2007 a third program under the acronym *STARS* was started (*Science and Transcendence Advanced Research Series*). Small groups of scientists and scholars from the humanities were to demonstrate how in the light of philosophic and theological considerations the natural sciences point to the nature, character, and significance of an ultimate reality. Twenty research units were endowed with 20,000 – 320,000 US dollars so that the whole program comprised US$ 1.3 million. In his essay Robert Russell reports on these undertakings and names the institutions which participated in them and the topics they treated.

In terms of research *CTNS* has collaborated since 1987 with the Vatican Observatory on a project called "Scientific Perspectives on God's Activity" initially estimated to last fifteen years. To advance this project scientific congresses have been organized in various places. There is also the J. K. Russell Fellowship which annually invites significant persons from theology or the natural sciences to teach in Berkley, to present public lectures, and to engage in dialogue with those interested. Invitees were, among others, Celia Deane-Drummond, Nancey Murphy, Wolfhart Pannenberg, Paul Davies, and John Polkinghorne.

With regard to teaching, the faculty of *CTNS* is present on the campus of the Graduate Theological Union (*GTU*) in Berkley to offer seminars and lecture courses at various theological in-

---

[20] Cf. Robert John Russell, "STARS: Science and Transcendence Advanced Research Series – Completing the Grant, Continuing the Research", *Theology and Science* (2010/4), 8:347 – 355, for the following.

stitutions. Ted Peters (*1941), Distinguished Research Professor of Systematic Theology at Pacific Lutheran Theological Seminary, member of the Core Doctoral Faculty at the Graduate Theological Union, has been involved in many ventures of *CTNS* as program or project director or in other important functions. – Robert Russell is however the only fulltime employed faculty member. – With the help of the *Templeton Foundation*, *CTNS* sees as one of its important tasks beyond what it does in Berkley, the promulgation of events in the academic arena on the worldwide scale and also to develop appropriate teaching aids. For instance, numerous books have been published by scholars who are closely associated with *CTNS* or were awarded a *CTNS* fellowship. Finally the publications of *CTNS* itself must be noted. They include first of all the documentary volumes of the joint research program with the Vatican Observatory, the quarterly journal *Theology and Science*, of which Russell and Peters are the editors, and *The CTNS Bulletin* in which all pertinent information about *CTNS* appears.

The scientific emphasis of *CTNS* is in the area of physics, cosmology, evolutionary biology, and genetics and secondarily in the neural sciences, technology, environmental research, and mathematics. Central in its endeavor is the dialogue with Christian theology, ethics, and spirituality, as well as, in more recent years, issues which result from the encounter between the natural sciences and the world religions. Through the untiring engagement of Robert Russell *CTNS* has become an extremely fruitful place for the dialogue between theology and the natural sciences. While *CTNS* conducts the dialogue from the perspective of a moderately conservative mainline theology one should not omit the considerable number of more conservative scholars.

Note must also be made of the aforementioned *Metanexus Institute*. Its beginnings were in 1997 with William Grassie then at Temple University who started a network on science and religion. Together with four faculty members of the University of Pennsylvania he founded what became the *Philadelphia Center on Religion and Science* (*PCRS*) and since 2000 the *Metanexus Institute*. It has hosted numerous national and international conferences and since 2011 has moved its offices from Philadelphia to New York. For its various conference projects it has received sizable grants from the *Templeton Foundation*. Some of its re-search projects are "spiritual transformation," "the unity of knowledge," and "subject, self, and soul." The institute is also involved in book publishing. Metanexus fosters a growing international network of individuals and groups exploring the dynamic interface between cosmos, nature and culture.

Among the conservatives the name of BERNARD RAMM (1916–1992), a Baptist whose book *The Christian View of Science and Scripture* (1954) was quite influential for evangelical theology, must be mentioned. While he betrays in the book almost a biblicist approach, he does not explain in literalistic terms the Genesis narrative of the creation of the world and the flood. He also exerted considerable influence on the *American Scientific Affiliation (ASA)* so that it did succumb to literalism in contrast to the creationists who in the 1960's propagated a literalistic approach to the Bible. Ramm emphasized that biblical revelation did not anticipate any scientific discoveries even if both occasionally agree with each other. The Holy Spirit did not divulge to the biblical writers the secrets of modern science but expressed infallibly in words true theological doctrines albeit in the cultural conceptuality of their respective times.[21] Ramm was an apologetic theologian, influenced by Karl Barth under whom he studied in Switzerland. In contrast to Barth he was very much interested in asserting the truthfulness of the biblical message in face of modernity whereby his starting point was the attestation of the truthfulness of Scripture by the Holy Spirit through the faithful. In 1979 the journal of the *American Scientific Affiliation* devoted a special issue to his book *The Christian View of Science and Scripture* commemorating the 25th anniversary of its publication, calling it "A Bernard Ramm Festschrift".[22] The Old Testament scholar C. JOHN COLLINS (*1954) appropriately called this book "a classic from a conservative evangelical author."[23]

---

[21] Cf. Bernard Ramm, *The Christian View of Science and Scripture* (Grand Rapids, MI: Wm. B. Eerdmans, 1954), 136.

[22] *Journal of the American Scientific Affiliation*, vol. 31 (December 1979).

[23] C. John Collins, *Science & Faith: Friends or Foes?* (Wheaton, IL: Crossway, 2003), 351.

With Ramm we have also come to the *American Scientific Affiliation* which was founded in the USA in 1941 with the goal of attaining a positive relationship between the sciences and the Christian faith. The *ASA* deals with every area and issue which pertains to the natural sciences and the Christian faith and presents to the Christian and scientific communities the results of such investigations for critique. Since the expansion of the natural sciences often resulted in a reaction of fear among the faithful the *ASA* wants to show that true theology and true natural science do not threaten each other but that there is an innate harmony between the two.

The *ASA* is organized regionally as well as nationally and conducts annual meetings. Through circular letters it maintains contact with its membership and also through a quarterly journal. One can become a member if one has a degree in science. By their standards science comprises everything from anthropology to the so-called hard sciences and also sociology. Those who have no degree in the natural sciences but agree to the statement of faith of the association can become associate members. This statement of faith says that "*the Holy Scriptures are the inspired Word of God, the only unerring guide for faith and conduct.*" As God's administrator one has responsibility for creation to use science and technology for the good of humanity and the whole world. "*God is the Creator of the physical universe. Certain laws are discernible in the manner in which God upholds the universe. The scientific approach is capable of giving reliable information about the natural world.*" Though the association is evangelical in persuasion it propagates neither fundamentalism nor creationism.

RALPH WENDELL BURHOE (1911–1997) is from a different Baptist tradition than Ramm. Burhoe entered Andover Newton Theological School in 1932 with the intention of becoming a pastor. Yet gradually his interest changed and he found a position at the Blue Hill Meteorological Observatory of Harvard University where he studied meteorology and climatology as an undergraduate. From 1947–64 he was an executive officer of the *American Academy of Arts and Sciences (AAAS)*.

It was during his time at the *AAAS* that Burhoe was instrumental in founding the *Institute on Religion in an Age of*

Science (*IRAS*, 1955). This institute was intended to bring together the different departments of human understanding because "any doctrine of human salvation cannot successfully be separated from realities pictured by science."[24] Beyond that it sponsored the annual conference on "Religion in an Age of Science," or the Star Island Conferences held on Star Island off Portsmouth, New Hampshire.[25] There were also other conferences and seminars, a publication program, lectures and seminars at colleges and theological schools. Furthermore at the *Unitarian Meadville/Lombard Theological School* in Chicago a center for research and advanced studies was instituted. This means that Burhoe's vision of a new natural theology based on science was most receptively welcomed by the *Unitarian Universalist Association.*

In 1966 Burhoe founded *Zygon: Journal of Religion and Science.* That same year the *Center for Advanced Study in Theology and the Sciences* (*CASTS*) was also established at the Meadville/Lombard Theological School. In 1972 this center was renamed *Center for Advanced Study in Religion and Science* (*CASIRAS*). In 1972 Burhoe retired from Meadville, and in 1980 he received the Templeton Prize for Progress in Religion. In 1988 *CASIRAS* joined with the *Lutheran School of Theology at Chicago* (*LSTC*) to establish the *Chicago Center for Religion and Science* under the leadership of the theologian PHILIP HEFNER (*1932) and the physicist THOMAS GILBERT (*1922). The *Chicago Center for Religion and Science* was later renamed in the *Zygon Center for Religion and Science.* The present director of the *Center* is Lea F. Schweitz who also serves as Associate Professor of Systematic Theology/Religion and Science at *LSTC,* the institution which is now also the home for *Zygon* and for the Annual Chicago Ad-

---

[24] According to the "Statement of Purpose," reprinted in David R. Breed, "Ralph Wendell Burhoe I," *Zygon* (1990*)*, 25:348.

[25] This annual seminar now takes place at various other locations since Star Island is no longer available for this conference. *IRAS* also publishes *Zygon* and organizes seminars and lectures on the interchange between theology and the sciences.

vanced Seminar in Religion and Science that Burhoe had began at
Meadville in 1966.

Burhoe considered religion from a socio-evolutionary perspective as "an evolving cultural art whose function is to orient us
to the ultimate goals and conditions for life at the top of the hierarchy of values."[26] To attain that kind of orientation Burhoe
advocated a rational interpretation of religion in the light of the
sciences. In so doing we can mine the insights of the sciences to
enlighten the human mind and to further the central role of humanity for human welfare. The most important accomplishment
of Burhoe was to facilitate dialogue among theologians and scientists of often quite different ideological religious and theological persuasions, albeit within the context of the Judeo-
Christian tradition. His journal *Zygon*, similar to *IRAS* often with
a very liberal persuasion, achieves a circulation far beyond the
confines of theological journals. From 1990 to 2008 *Zygon* was
edited by Philip Hefner and since then by the Dutch philosopher
of religion WILLEM B. DREES (*1954).

6.2.3 British Institutions

As we turn now to Great Britain we notice there the existence of
some honorable lecture series such as the Gifford Lectures which
have been conducted since 1888 by the Scottish universities St.
Andrews, Glasgow, Aberdeen, and Edinburgh. There was only a
brief interruption caused by World War II. They date back to the
lawyer and judge LORD GIFFORD (1820 – 1887) who willed to these
universities 80,000 pounds to "promote and diffuse the study of
natural theology in the widest sense of the term—in other words,
the knowledge of God." Natural theology should thereby be
treated as a science "without regard or trust to any assumed
special or so-called supernatural revelation." With this emphasis
on natural theology we can clearly discern the legacy of the Enlightenment. Therefore it is a matter of course that famous scientists presented lectures in this series, such as Werner Heisen-

---

[26] So David R. Breed, "Ralph Wendell Burhoe V," *Zygon* (1991), 26:411.

berg (1955/56) or Niels Bohr (1949/50). Yet when we notice that even Karl Barth (1936–38), Reinhold Niebuhr (1938–40), and Rudolf Bultmann (1954/55) were invited to present lectures one realizes that until the mid-1960's scholars who concern themselves with the dialogue between theology and the natural sciences were in this series more the exception than the rule.

The widespread interest in the dialogue between theology and the natural sciences did not gain momentum until the second half of the 20th century. In 1985, for instance, the *Ian Ramsey Center* was founded in Oxford which deals with issues that result from the interface between theology and science. For many years the center was also financially supported by the *Templeton Foundation*. – In 1990 the Greek medical doctor ANDREAS IDREOS (1917–1997) endowed at Oxford University a professorship for science and religion which from 1999 to 2006 was occupied for the first time by the historian of science JOHN HEDLEY BROOKE (*1944). His successor was the philosopher of religion Peter Harrison (*1955). Here too the *Templeton Foundation* is supporting the professorship with an annual department lectureship in science and religion and three fully funded three-year doctoral scholarships. – In order to bring together the fields of science and religion the author and novelist SUSAN HOWATCH (*1940) donated a part of the revenues from her novels to finance an academic position of a "Starbridge Lecturer in Natural Science and Theology" at the theological faculty at Cambridge University. The psychologist and theologian Fraser N. Watts was the first lecturer to be called to this position in 1994. – At the University of Chester in 2002 the *Center for Religion and the Biosciences* was established at which Celia Deane-Drummond occupied the chair in theology and the biological sciences from 2002–2011. Once the center was erected she became the director of that center. In 2011 she accepted a full professorship in theology at the University of Notre Dame in Indiana, USA and there has a unique appointment in conjunction with the Department of Theology, the College of Arts and Letters, and the College of Science. This shows again the vivid interest in an exchange between theology and the sciences.

On the Evangelical side there is the *Victoria Institute*, founded in 1865 with a long tradition of wanting to bring together the

Christian faith with the progress in scientific knowledge. Together with the *Christians in Science* which is a member of the
Evangelical Alliance the *Victoria Institute* issues the journal *Science and Christian Belief* which appears twice a year. Its first
edition was published in 1989. Furthermore there is a more
popular journal *Faith and Thought* which wants to communicate
to non-specialists the pastoral and ethical consequences of scientific progress. The aforementioned *Christians in Science* is an
international network. Membership is open to scientists, teach-
ers, students and all others who have an interest in the dialogue
between the sciences and the Christian faith.

In our survey we quickly noticed that in many countries the
institutionalization of the dialogue between theology and the
sciences has made great progress in the last few decades. But in
Germany at least the universities as the most important educational institutions have not yet realized this need for institutionalization. DIRK EVERS (*1962) for instance, is a member
of the Council of *ESSSAT*. But at the University of Halle-Witten-
berg he holds a full professorship and deals with the areas of
systematic theology, practical theology, and science of religion.
This is just one case among many. Therefore in Germany and in
many other countries the dialogue is restricted to the initiative of
individual persons who are engaged in it, to occasional conferences in the academies of the Protestant and Roman Catholic
churches, and to meetings of associations supported largely by
their members.

The best dialogue does not amount to much if nobody hears
about it. Therefore it is important that publishing houses dedi-
cate special series to this dialogue. Fortress Press in Minneapolis, USA, has with *Theology and the Sciences* most likely the
oldest series with more than thirty publications. On its editorial
board are well-known theologians such as Robert Russell, Sallie
McFague, and John Polkinghorne. On the British side we should
mention the Ashgate *Science and Religion Series* which is
managed by Roger Trigg and Wentzel van Huyssteen. In Germany *Religion, Theology, and Natural Science: Religion, Theologie und Naturwissenschaft* by the publisher Vandenhoeck &
Ruprecht in Göttingen has grown to more than twenty-five

volumes.[27] On its editorial board are Ted Peters, Antje Jackelèn, and Willem Drees.

---

[27] One day I was surprised by a telephone call to my home in Germany from Ian Barbour in which he asked: "Hans, is Vandenhoeck & Ruprecht a respectable publisher, since they want to publish my Gifford Lectures?" In good conscience I could say "Yes". Then they published, as the first volume in their series, his book *Wissenschaft und Glaube* (2003: *Religion in an Age of Science,* HarperSanFrancisco, 1990).

7. Partners from the Natural Sciences

After this long excursion into the history and institutionalization of the dialogue between theology and the natural sciences we now turn to the most important dialogue partners of the present. We begin with natural scientists and want to omit prestigious figures who have already died, such as the Australian physiologist and Nobel laureate SIR JOHN ECCLES (1903–1997), who asked in a memorable lecture: "What happens to the conscious self at brain death? … Is the self renewed in some other guise and existence? This is a problem beyond science, and scientists should refrain from giving definite negative answers."[1] We will also pass over the German physicist and philosopher Carl Friedrich von Weizsäcker of whom the former Saxon Prime Minister Kurt Biedenkopf said: "As a man of 'both worlds', the natural science and the human-ities, he has exemplified in his own life the dialogue between the natural sciences and the humanities."[2] Among the theologians too we will confine ourselves to the present-day dialogue partners and therefore omit Langdon Gilkey who in 1970 in a noteworthy publication *Religion and the Scientific Future* emphasized that "the vast new powers of science do not, in the end, make religious faith and commitment irrelevant; they make them more neces-

---

[1] John C. Eccles, in his lecture, "The Brain-Mind Problem as a Frontier of Science," in *The Future of Science: 1975 Nobel Conference*, 88.

[2] Kurt H. Biedenkopf, "Das Recht auf Utopie," in Carl Friedrich von Weizsäcker, *Die Zeit drängt / Das Ende der Geduld: Aufruf und Diskussion* (Munich: dtv, 1989), 153.

sary than ever."[3] In the 1980's he vehemently objected to the newly flourishing creationism and its claim that it would offer a scientific worldview.[4] But who are today the most important dialogue partners?

7.1 Discerning the Mind of God (Stoeger, Davies, Hawking, Tippler)

The natural sciences investigate the world and attempt to discern the way by which things are or why events occur. This curiosity is constitutive for science and indispensible for scientific progress. Given this mentality it is not surprising that many scientists who engage in the dialogue with theology have on their agenda to discern the mind of God whether they openly admit this or pursue it implicitly. This is perhaps already the idea behind the collaboration between the Vatican Observatory and the *Center for Theology and the Natural Sciences* in Berkeley.

WILLIAM R. STOEGER, S. J. (*1943) first studied philosophy and theology and then received his doctorate from the University of Cambridge in astrophysics. From 1979 he worked at the Vatican Observatory and in 1990 he went from there with a Vatican research group to the Stewart Observatory of the University of Arizona where he also teaches as an Adjunct Associate Professor of Astronomy. Under his leadership the Vatican Observatory started the above mentioned collaboration with the *CTNS* in Berkeley resulting in a series of interdisciplinary conferences to discuss the issue of divine activity from scientific, philosophic, and theological perspectives. In a 1992 conference, for instance, the topic was "quantum cosmology and the natural laws", in 1994 "chaos and complexity", and in 1996 "evolutionary and molecular biology", etc. The latest publications from these conferences deal with the

---

[3] Langdon Gilkey, *Religion and the Scientific Future: Reflections on Myth, Science, and Theology*, 98.

[4] Cf. Langdon Gilkey, *Creationism on Trial: Evolution and God at Little Rock* (Charlottesville: University of Virginia, 1985).

biological roots of evil.[5] Along with his own proper research and publications in cosmology and astrophysics Stoeger ventured, through his collaboration with *CTNS*, to interdisciplinary issues. This is evidenced already when with regard to the universe he says that first there is the observable universe accessible to us, then the universe as a whole to which we have no direct access but whose existence and some of its characteristics can be calculated from the observable universe, and then the whole which is either congruent with the universe as a whole or much larger. This last one is withdrawn from scientific description.[6]

PAUL DAVIES (*1946) is a British physicist and followed the dialogue between theology and the sciences in his own way. He received his doctorate in the field of electrodynamics and worked for two years at the University of Cambridge with FRED HOYLE (1915 – 2001), who became especially known through the Steady State Theory. – According to this theory the universe expands steadily with uniform velocity through the continuous production of matter. Therefore there was not an explosive Big Bang. This theory however, Hoyle later abandoned as scientifically untenable. – Davies first taught in England, then in Australia, and since 2006 at Arizona State University in the USA. Besides cosmology and quantum theory Davies concerns himself with astrobiology and the issue of the origin of life. In wider circles he has been known through his generally very accessible books in which he deals with issues of worldview and of religion. In 1995 he received the Templeton Prize. He has written approximately twenty popular scientific books which have been translated into several languages. Besides the Templeton Prize Davies also received many other honors, for instance in 2002 the Michael Faraday Prize of the Royal Society of London, Great Britain.

---

[5] *Physics and Cosmology: Scientific Perspective on the Problem of Natural Evil*, vol. 1, Nancey Murphy/Robert John Russell/William Stoeger, S. J., eds. (Notre Dame, IN: University of Notre Dame, 2007).

[6] Cf. William R. Stoeger, "What is 'the Universe' which Cosmology Studies?" in *Fifty Years in Science and Religion. Ian Barbour and His Legacy*, Robert H. Russell, ed. (Burlington, VT: Ashgate, 2004), 141 f.

When Paul Davies received the Templeton Prize he said among other things:

> Some scientists have tried to argue that if only we knew enough about the laws of physics, if we were to discover a final theory that united all the fundamental forces and particles of nature into a single mathematical scheme, then we will find that this superlaw, or theory of everything, will describe the only logically consistent world. In other words, the nature of the physical world would be entirely a consequence of logical and mathematical necessity. There would be no choice about it. I think this is demonstrably wrong. There is not a shred of evidence that the universe is logically necessary. Indeed, as a theoretical physicist I find it rather easy to imagine alternative universes that are logically consistent, and therefore equal contenders for reality.[7]

Similarly to Stoeger, Davies claims that the world cannot be pressed into a mathematical concept though which one could describe the ultimate reality. One could conceive of the world quite differently from what one thinks possible in logical and mathematical terms. For instance, he can justifiably claim that the ten years of radio-astronomy have taught humanity more about creation and the organization of the universe than thousands of years of religion and philosophy. Yet Davies does not want to limit God to a first cause which at one time started the universe, because this is a much too anthropomorphic view of God.[8]

Modern quantum theory often seems to do away with the notion that everything needs to have a cause and that this first cause is then identified with God.[9] But Davies is uncomfortable with this view. God or some prime cause is still needed to institute the laws that govern the universe. This is an idea that Davies pursues more vigorously in a later publication, *The Mind of God.* God is not the one who initially pushed the button to get everything rolling, but he is the "universal mind

---

[7] *http://cosmos.asu.edu/prize_address.htm* (August 2011).

[8] Cf. Paul Davies, *Space and Time in the Modern Universe* (Cambridge: University Press, 1977), 217.

[9] Cf. Paul Davies, *God and the New Physics* (New York: Simon & Schuster, 1983), 42 f. and 216 f.

existing as part of that unique physical universe."[10] This physical universe is the medium through which God's spirit is expressed. Instead of a supernatural God, Davies claims to opt for a natural one, one that seems to have its affinity in the God of process thought.[11] At the same time Davies is cautious about granting too much insight to the natural sciences. By investigating the workings of nature, a scientist cannot discern anything about God's plan for the world or about the battle between good and evil. Science is reductionistic, and this is its main contribution to our knowledge of the world. Physics, for instance, does not deal with the questions of the directedness of the creative process or with morals.[12] Therefore religion, according to Davies, is not on the way out.

Whether we are scientists or not, we still search for a deeper sense of life. "Many ordinary people too, searching for a deeper meaning behind their lives, find their beliefs about the world very much in tune with the new physics."[13] Our worldview has changed so dramatically that the biblical worldview seems out of touch with the way we perceive the world today. This is one of the reasons why theologians should listen to science. It may provide them with the raw materials for the reconstruction of religious views. Davies thinks this reconstruction is quite advantageous, because he is "fully committed to the scientific method of investigating the world" since "science leads us in the direction of reliable knowledge."[14] While he does not subscribe to "conventional religion," Davies has come to believe more and more strongly, inspired by his scientific work, "that the physical

---

[10] Davies, *New Physics*, 223.

[11] Cf. Paul Davies, *The Mind of God: The Scientific Basis for a Rational World* (New York: Simon & Schuster, Touchstone Book, 1992), 183, where he admits his affinity to Whitehead, who "replaces the monarchical image of God as omnipotent creator and ruler to that of a participator in the creative process."

[12] So Davies, *New Physics*, 229.

[13] Davies, *New Physics*, vii.

[14] Davies, *The Mind of God*, 14. The parenthetical numbers in the following text refer to pages in *The Mind of God*.

universe is put together with an ingenuity so astonishing" that he "cannot accept it merely as a brute fact" (16). Moreover, he is convinced that sooner or later we all have to accept something as given, whether it is God, logic, a set of laws, or some other premise for our existence. The ultimate questions will always lie beyond the scope of empirical science, since it is there that science and logic will fail us.

Contrary to Einstein, Davies asserts that "God plays dice with the universe" (191). The statistical character of atomic events and the instability of many physical systems with minute fluctuations make it possible for the future to remain open and undetermined by the present. This allows new forms and systems to emerge so that the universe is endowed with freedom for genuine novelty. At the same time, however, Davies shows great admiration for the ontological and cosmological arguments for the existence of God. He is aware that the design argument has been resurrected in recent years by a number of scientists, and he points to the long list of "lucky accidents" and "coincidences" which have allowed life as we know it to evolve (cf. 199). In line with the anthropic principle, he toys with the idea that "the apparent 'fine-tuning' of the laws of nature necessary if conscious life is to evolve in the universe then carries the clear implication that God has designed the universe so as to permit such life and consciousness to emerge. It would mean that our own existence in the universe formed a central part of God's plan" (213).

At the same time, he knows that a decision for such a "designer universe" cannot be based on strictly scientific judgment but "is largely a matter of taste" (220). So he concludes with the confession: "I cannot believe that our existence in this universe is a mere quirk of fate, an accident of history, an incidental blip in the great cosmic drama. . . . We are truly meant to be here" (232). Since in the preface of *The Mind of God* he stated that his book is "more of a personal quest for understanding," such a confession at the conclusion is not inappropriate.

The philosopher LUDWIG WITTGENSTEIN (1889 – 1951) confessed at the end of his *Tractatus Logico-Philosophicus:* "We feel that even when all possible scientific questions have been answered, the problems of life remain completely untouched."

Wittgenstein went on to refer to the mystical.[15] In like manner
Paul Davies realizes that there must be something beyond science. He feels that scientific evidence may be interpreted to point
beyond science, though this is not a necessary step. With this kind
of hesitancy, Davies pushes scientific scrutiny and investigation
to its outermost limits while at the same time distinguishing
himself from scientists such as Stephen Hawking and Frank Tipler who have no qualms about going beyond that which seems
scientifically warranted.

STEPHEN HAWKING (*1942), held the post of Lucasian professor
of mathematics from 1979–2009 at the Department of Applied
Mathematics and Theoretical Physics at Cambridge University.
Presently he is director of research there at the Center for Theoretical Cosmology at the same university. Already in 1978 *Time*
magazine named Hawking "one of the premier scientific theo-
rists of the century, perhaps an equal of Einstein."[16] He focuses
exactly on those issues which, according to the traditional understanding of the natural sciences, are left out in scientific discourse. As he freely admits, he was motivated in his research in
cosmology and quantum theory by questions such as these:
"Where did the universe come from? How and why did it begin?
Will it have an end, and if so, how?"[17] He rightly claims that these
issues are of interest to all of us. To circulate the results of his
research to a wider audience, he wrote an informative and popular small book with the title *A Brief History of Time*. He concedes
that in 1970 he still thought "that there must have been a big bang
singularity" (50). This thesis had won general acceptance. But
now he feels that it is perhaps ironic that he has changed his mind

---

[15] Ludwig Wittgenstein, *Tractatus Logico-Philosophicus*, trans. D. F.
Pears/B. F. McGuinness, intro. Bertrand Russell (London: Routledge &
Kegan Paul, 1961), 149 (6.52).

[16] According to John Boslough, *Stephen Hawking's Universe* (New York:
William Morrow, 1985), 59, in his readable and informative biography.

[17] Stephen W. Hawking, *A Brief History of Time: From Big Bang to Black
Holes*, intro. Carl Sagan (London: Bantam Books, 1988), vi. The parenthetical numbers in the following text refer to pages in this work.

and attempts to persuade other physicists to accept exactly the opposite, namely, that in the beginning of the universe there was no such singularity.

Hawking starts with the conviction that natural science has uncovered those laws that, within the limits of the Heisenberg uncertainty principle, tell us how the universe will develop within time, provided we know its state at any one time. According to Hawking, these laws may have initially been decreed by God, but then he seems to have left the universe and no longer intervenes. Yet why has God chosen this original state or configuration of the universe?

> What were the "boundary conditions" at the beginning of time? One possible answer is to say that God chose the initial configuration of the universe for reasons that we cannot hope to understand. This would certainly have been within the power of an omnipotent being, but if he had started it off in such an incomprehensible way, why did he choose to let it evolve according to laws that we could understand? The whole history of science has been the gradual realization that events do not happen in an arbitrary manner, but that they reflect a certain underlying order, which may or may not be divinely inspired. (122)

Hawking wrestles here with the problem that we understand how the universe developed fairly well, but we do not know how this development came about. He surmises that it is very difficult to suppose that from an initial chaotic configuration a universe emerged which in large areas was as uniform as we know it today. To resolve the apparent contradiction between an initial chaos and the subsequent order, Hawking introduces an imaginary time. Imaginary numbers (i) are numbers which when multiplied with themselves still yield negative values, for instance, $2i \times 2i = -4$. By connecting our Euclidean understanding of space and time with the quantum theory of gravity, he arrives at the possibility in which there would be no boundary to space-time and so there would be no need to specify the behavior at the boundary. There would be no singularities at which the laws of science broke down and no edge of space-time at which one would have to appeal to God or to some new law to set the boundary conditions for space-time. One could simply say: "The boundary condition of the universe is that it has no boundary." The universe would be completely self-contained and not affected by anything outside itself. It would neither be created nor destroyed. It would just BE. (136)

Seen from our real time, we could still assert that our universe was at its smallest 10 or 20 billion years ago, which would correspond to the maximum radius of history in imaginary time (138). Later, in analogy to a chaotic inflation model, our universe would have expanded. After this phase of expansion it would again slowly contract. Our view of the world would not be that different from the way it used to be. But in relation to imaginary time there were no singularities, neither at the beginning nor at the end.

Hawking knows that the idea of space and time as a surface closed in itself without boundaries has far-reaching consequences for our understanding of how God is related to the universe (for the following, including the quote, see 140 f.). Up to now one could still think that God had given the universe certain laws according to which it develops and that through divine intervention God could break through these laws. Since these laws do not tell us what the universe looked like at the beginning, "it would still be up to God to wind up the clockwork and to choose how to start it off. So long as the universe had a beginning, we could suppose it had a creator. But if the universe is really completely self-contained, having no boundary or edge, it would have neither beginning nor end: It would simply be. What place, then, for a creator?"

Hawking says in conclusion that scientists thus far have been so busy developing new theories to describe the "what" of the universe that they have had no time to ask the why question (for the following, including the quote, see 174 f.). On the other hand, philosophers who traditionally have posed the why question had no time to keep pace with the progress in the natural sciences. Therefore, according to Hawking, they have confined themselves more and more to an analysis of language. But once a complete and comprehensive theory is developed, philosophers, scientists, and all other people must participate in the discussion of the question "of why it is that we and the universe exist. If we find the answer to that, it would be the ultimate triumph of human reason — for then we would know the mind of God."

In *A Briefer History of Time* (2005) a popular and more recent edition of the abovementioned publication Hawking admits on the one hand that a theory which explains everything can never be proven because a theory remains a theory even if it is well-founded. Nevertheless it would "revolutionize the ordinary person's understanding of the laws that govern the universe" even if it were just the first step on the way to a complete understanding of the events around us and of our existence.[18] Important here is the admission that a theory can always be falsified and essentially demonstrate its actual truthfulness only at the end of history. On the other hand Hawking is convinced that in unearthing that which the world holds together in its core the question of God becomes superfluous because then the facts of all secrets are laid open.

In the tradition of William Paley, Hawking compares the function of God with that of a watchmaker. And in analogy to the classical Thomistic understanding of miracles, he understands the present activity of God as an intervention, an interruption or a breakthrough of the ruling laws of nature. This shows in a paradigmatic way that Hawking, similar to many other scientists, has an understanding of God that corresponds neither to biblical understanding nor to current theological formulations. Since Hawking, however, is a natural scientist, one cannot fault him for that oversight. Much more important are two other points, namely that he clearly realizes that scientists today can no longer ignore the why-question. A simple description of 'what is', or 'how it happens', does not suffice. Especially from an ecological perspective, it has become increasingly urgent to include value judgments in our considerations. If we want to find direction in today's world and gain meaning for our existence we can neither be satisfied with a description of the state of the world and the cosmos nor of its trends.

In line with the British skeptic tradition Stephen Hawking resolved the why-question in his own way. There is no mysterious initial singularity, but the states prior to and after the "beginning" are merged into each other. The universe is closed in itself. If we were to adopt this view, we would return to the closed causal-mechanistic worldview of the 19th century. Yet at the end Hawking

---

[18] Stephen Hawking, *A Briefer History of Time*, with Leonard Mlodinow (London: Bantam, 2005), 136.

admits that the why-question is still not resolved. He still hopes for a complete theory of everything as a future possibility. To stay with Hawking's terminology, we still cannot understand the mind of God. For Hawking this is not a mystery which cannot be solved in principle, but rather one that we can perhaps solve.[19] Yet he cannot adduce any justified reason why such a hope would find its fulfillment. We are confronted here with a basic question of why something is and why there is not simply nothing. Hawking has pushed back one step further the vexing questions concerning creation but he has not solved them. Theologically speaking this should not surprise us. Hawking is motivated in his research by why-questions. In pursuing them he necessarily comes up against the limits of that which science can ascertain.

FRANK J. TIPLER (*1947), professor of mathematical physics at Tulane University in New Orleans, wrote *The Anthropic Cosmological Principle* with John D. Barrow (*1952). In this book he wanted to introduce a new teleology on a scientific basis. Then he wrote the voluminous book *The Physics of Immortality,* in which he picks up on classical American Deism and transforms it to a rational Christian faith. In unfolding the anthropic principle, Tipler first rejects the fundamental presupposition that humanity as an observer in the universe should not have a privileged position. Instead he shows that an observer can only occupy the place where the conditions are right for his evolution and existence so that he can emerge. Such a place must necessarily be something special. Tipler refers to the thesis of the British theoretical physicist BRANDON CARTER (*1942), first introduced in 1974, in which Carter already described the anthropic principle. Carter claimed "that what we can expect to observe must be restricted by conditions necessary for our presence as observers. (Although our situation is not necessarily *central,* it is inevitably privileged to some extent.)"[20]

---

[19] Cf. to this point Ian G. Barbour, *Religion in an Age of Science: The Gifford Lectures 1989–1991* (San Francisco: Harper, 1990), 1:140.

[20] Brandon Carter, "Large Number Coincidences and the Anthropic

Tipler and Barrow start with the statement that there are a number of very unlikely — and coincidental — accidents which are totally independent from each other. These accidents seem to be necessary if an observer, whose basic building material is carbon compounds, is to emerge in our universe.[21] This leads them to establish three different anthropic principles. First there is the weak anthropic principle which states: *"The observed value of all physical and cosmological quantities are not equally probable, but they take on values restricted by the requirement that there exist sites where carbon-based life can evolve and by the requirement that the Universe be old enough for it to have already done so"* (16). The strong anthropic principle then goes one step further and claims: *"The Universe must have those properties which allow life to develop within it at some stage in its history"* (21). Lastly, they formulate the final anthropic principle, which Tipler has expanded in his second book. It says: *"Intelligent information-processing must come into existence in the Universe, and, once it comes into existence, it will never die out"* (23). How far Tipler and Barrow go beyond the strict limits of physics at this point becomes visible when they claim on the basis of the strong anthropic principle that it is incomprehensible why life should become extinct once it has evolved but has not yet decisively influenced the universe at large. Tipler and Barrow emphasize the necessity of a human observer. In classical physics, humanity had an ancillary role in the universe. But in modern physics, especially in the Copenhagen interpretation of quantum mechanics, the human observer plays a very decisive role (cf. 458).

Two other things are important for Tipler and Barrow: First, the possibility of space travel. Humanity is no longer a fringe phenomenon in the immense depth of the universe. At least in principle humanity can gradually colonize essential parts, if not

---

Principle in Cosmology," in *Confrontation of Cosmological Theories with Observational Data,* ed. M. S. Longair (Dordrecht: D. Reidel, 1974), 291.

[21] Cf. John D. Barrow/Frank J. Tipler, *The Anthropic Cosmological Principle,* 2nd ed. (Oxford: Clarendon Press, 1988), 5, where they follow Carter's lead. The parenthetical numbers in the following text refer to pages in Barrow's and Tipler's work.

all, of the visible cosmos (cf. 613 f.). The second thing which has impressed Tipler and Barrow is computer technology. In analogy to that technology, humanity does not consist of body and soul but "a program designed to run on a particular hardware called a human body" (659). Such a program can then be duplicated and transferred at will. Having stated this, the authors conclude the anthropic principle with a similar scenario, as Tipler's more recent book, *The Physics of Immortality,* shows. The final point of development, meaning the omega point, is attained once "life will have gained control of *all* matter and forces not only in a single universe, but in all universes whose existence is logically possible; life will have spread into *all* spatial regions in all universes which could logically exist, and will have stored an infinite amount of information, including *all* bits of knowledge which is logically possible to know. And this is the end" (677).

While the weak anthropic principle only states the obvious, namely, that only under certain conditions can life exist and that without these conditions it could not have existed, the strong and the final anthropic principles start with presuppositions that are not without problems.[22] The strong anthropic principle presupposes that certain initial data must exist, while the final anthropic principle is a mere postulate on the basis of the strong anthropic principle. – In a more recent book Barrow calls the "'Weak Anthropic' consideration a methodological principle which, if ignored, will lead one to draw incorrect conclusions from the data at hand."[23] It takes more than ten billion years for nature to produce the building blocks of life. Therefore a universe as huge and as old as ours was needed so that human life could evolve. Since "many constants of Nature owe their values to quasi-random processes occurring in the earliest stages of the Universe," our universe could have taken a very different shape from what we see today.[24] There was no necessity that our universe evolved the way it actually did.

---

[22] Cf. also the objections made by Willem B. Drees, *Beyond the Big Bang: Quantum Cosmologies and God* (La Salle, Ill.: Open Court, 1990), 78 f.

[23] John D. Barrow, *New Theories of Everything; The Quest for Ultimate Explanation* (Oxford: Oxford University, 2007), 195.

[24] Barrow, *New Theories of Everything,* 197.

He rightly cautions that, because of the uncertainties involved, "there is no formula that can deliver all truth."[25]

In *The Physics of Immortality,* a highly controversial book simultaneously published in the USA and Germany, Tipler, however, has given up any theological restraints and immediately claims in the preface that "physicists can infer by calculation the existence of God and the likelihood of the resurrection of the dead to eternal life in exactly the same way as physicists calculate the properties of the electron."[26] He sees his venture into theological realm justified, since in the twentieth century theologians have largely withdrawn from questions about nature and the cosmos. Wolfhart Pannenberg is "a very rare exception" because he is "one of the very few modern theologians to truly believe that physics must be intertwined with theology" (xxiiif.). But Tipler immediately reveals his reductionistic inclinations when he emphasizes: "It is necessary to regard all forms of life—including human beings—as subject to the same laws of physics as electrons and atoms. I therefore regard a human being as nothing but a particular type of machine, the human brain as nothing but an information processing device, the human soul as nothing but a program being run on a computer called the brain" (xi).

In *The Physics of Immortality,* Tipler describes his omega point theory as "a testable physical theory for an omnipresent, omniscient, omnipotent God who will one day in the far future resurrect every single one of us to life forever in an abode which is in all essentials the Judeo-Christian Heaven" (1). Such proof is indeed an immense program for a physicist. Tipler admits, however, that his omega point theory is "a viable scientific theory of the future of the physical universe, but the only evidence in its favor at the moment is the theoretical beauty, for there is as yet no confirming experimental evidence for it" (305). Tipler believes that the odds are quite high that the omega point theory is true. He calls himself neither a Deist, meaning, according to him, that

---

[25] Barrow, *New Theories of Everything,* 246.

[26] So Frank J. Tipler, *The Physics of Immortality: Modern Cosmology, God, and the Resurrection of the Dead* (New York: Doubleday, 1994), ix. The parenthetical numbers in the following text refer to pages in this work.

he does not believe in a personal creator God who affects the world from outside it, nor a Christian, since he neither accepts the resurrection of the dead nor the Trinity nor the real presence of Christ in the Lord's Supper. More appealing to him is American Deism, though he admits that its God is too impersonal. According to Tipler, "religion can be based on physics only if the physics shows that God *has* to be personal, and further, that the afterlife is an absolutely solid consequence of the physics" (327).

Tipler's main goal is to bridge the ugly, broad ditch between physics and religion. He considers it a grave problem that, for many theologians and natural scientists, religion and science have virtually nothing to do with each other, that religion concerns itself primarily with moral issues and science with facts. But according to Tipler, moral decisions, too, must be secured by facts. "If religion is permanently separated from science, then it is permanently separated from humanity and all of humanity's concerns. Thus separated, it will disappear" (332). Tipler therefore opts for an integration of theology into physics. He treats theological assertions as purely scientific assertions so that theology becomes for him "a branch of astronomy," and like any other science it is based only on reason and no longer on revelation (336). "Theology is nothing but physical cosmology based on the assumption that life as a whole is immortal" (for this and the following quote, see 338 f.). Science can offer the same consolation in the face of death that religion once offered. "Religion is now part of science."

In a more recent publication *The Physics of Christianity* (2007) Tipler is less skeptical and admits: "I believe in the Cosmological Singularity which *is* God."[27] According to Tipler the reason for that is very simple because one can "prove" that the essential contents of faith such as the existence of God, the Trinity, the miracle of the star at Bethlehem all the way to the virgin birth and the bodily resurrection of Jesus and his parousia are all facts which do not contradict the insights of physics and therefore are "proven."[28]

---

[27] Frank J. Tipler, *The Physics of Christianity* (New York: Doubleday, 2007), 268.

[28] This "proof" is spelled-out by Tipler in *The Physics of Christianity.*

With Tipler it becomes more evident than with all the scientists thus far reviewed that he has not only arrived at the boundary of the natural sciences but has transgressed this boundary by a great distance. It is indeed questionable whether one must distinguish between knowledge of the world and knowledge of salvation if God ultimately is the creator of the world and the mediator of salvation. Wolfhart Pannenberg, for instance, rejected the distinction between salvation history and world history. However if Einstein is correct that the parameters for orientation in our world, meaning space, time, and matter, have no independent validity but only make sense in relation to each other then it is difficult to understand how we can obtain from the world itself guidelines of how to conduct ourselves and find direction for our life. The evolving universal spirit which, according to Tipler, emerges at the end of the universe is nothing but that which potentially was already present at the beginning. At the most we can discern a factual but not a necessary direction in our world. The transition from the weak anthropic principle which actually is a tautology to the strong anthropic principle requires certain basic decisions which are not justified by the universe itself. This means Tipler is not a dispassionately calculating and descriptive scientist but is implicitly a theologian. He determines what is desirable for himself and what makes sense. But it is exactly here that the discussion with theology is needed because the unearthing the facts of this world by science cannot provide for such value judgments.

If we look around for authors in the German language medium it does not take long to notice that there are only a few. If we take for instance the notable work of Roman Catholic theologian ANDREAS BENK (*1957), *Moderne Physik und Theologie: Voraussetzungen und Perspektiven eines Dialogs* (*Modern Physics and Theology: Presuppositions and Perspectives of a Dialogue*; Mainz: Matthias-Grünewald, 2000) then we find the names of MAX PLANCK (1858–1947), Albert Einstein, and Werner Heisenberg mentioned, but no one from the present generation. This is also the case with Hans-Peter Dürr, *Physik und Transzendenz: Die großen Physiker unseres Jahrhunderts über ihre Begegnung* *mit dem Wunderbaren* (*Physics and Transcendence: The Great Physicist of Our Century in Their Encounter with the Miraculous*). After an extensive preface he renders only contributions of scientists from the past such as NIELS BOHR (1885–1962), Max Planck, or ERWIN SCHRÖDINGER (1887–1961). The paucity of a present-day engagement of the scientists in a dialogue seems to be founded in Dürr's claim: "Physics and transcendence in the conception of present-day physicists are no longer related to each other in an antagonistic way but rather in a complementary one."[29] This may be true for classical physics, but not for physics in the wider sense because even cosmology and biology proceed today largely on the basis of physics. Yet in both areas basic presuppositions and conclusions often are diametrically opposed to theological assumptions.

7.2 Between Rejection and Proof (Wuketits, Dawkins, Wilson, Kutschera, Gitt, Dembski, Scherer)

While for most serious scientists (and theologians) the metaphor of a warfare between theology and the natural sciences is totally inadequate, there is an increasing number of scientists of both liberal and extreme conservative ideological persuasion who engage in a crusade either against the alleged irreconcilability of theological claims of a creator and the facts of science or in the perceived Godlessness of science which destroys the Christian faith. This means some scientists assert that the facts of science render theological statements untenable while others attempt to show that the facts of science are different from what most believe and actually show beyond doubt that a creator is plausible.

For instance the Austrian biologist FRANZ M. Wuketits (*1955) who teaches in Vienna, Austria, advocates a strict atheism founded on evolutionary biology. He is a member of the scientific

---

[29] Hans-Peter Dürr, ed., *Physik und Transzendenz: Die großen Physiker unseres Jahrhunderts über ihre Begegnung mit dem Wunderbaren* (München: Scherz-Verlag, 1986), 11.

board of the *Giordano Bruno Foundation* which has as its goal "the advancement of an evolutionary humanism." According to Wuketits God is not a topic for the scientist because a scientist can neither prove nor disprove God's existence.

> Therefore he cannot take God as the starting point of his studies. ... A believer even when he or she accepts evolution as a fact can still maintain that the becoming of life, even the emergence of the cosmos is a symbol of the divine expression in this world; but he or she will then have to understand that with this assertion the realm of scientific research has been abandoned.[30]

Wuketits is correct when he strictly distinguishes between scientific and theological assertions.

In explaining the evolutionary theory of Darwin he emphasizes that natural selection as the most important factor in evolution excludes expediency in nature. Since selection is a mechanically working principle there remains no final goal in nature. "From this viewpoint there remains no place for faith in a cosmic teleology, a faith that all living beings are subordinated to a universal expediency."[31] Though Darwin's theory, as Wuketits shows, was considerably modified, in principle it has been proven accurate. Therefore Wuketits concludes: "Evolution as a *natural* process is today and remains also tomorrow the starting point of biology; any refuge into a metaphysical construct of thought or even into creationism would be a relapse into a way of thinking that is long past."[32] Of course it is true that biology regards all processes as natural, otherwise it would not be a science. Here we must agree with Wuketits.

But when he claims: "Evolution is the key to the understanding of everything living," he lapses into an absolutism which can no longer be substantiated.[33] Evolution is only one important theory

---

[30] Franz M. Wuketits, *Evolutionstheorien: Historische Voraussetzungen, Positionen, Kritik* (Darmstadt: Wissenschaft-liche Buchgesellschaft, 1988), 29.31.

[31] Wuketits, *Evolutionstheorien*, 50.

[32] Wuketits, *Evolutionstheorien*, 171.

[33] Wuketits, *Evolutionstheorien*, 173.

toward understanding everything living, but it is not the only theory. However, Wuketits is especially interested in such absolutes. Since for Wuketits evolution is the key to everything living he can discard the theodicy issue of how the evil of this world can be reconciled with a wise, benevolent, and almighty God. Already the fact that in the course of the history of the earth and of evolution approximately 500 million to 1 billion species have died out contradicts for him the possibility of an intelligent designer.[34] In contrast to a frequent interpretation of evolution and also in contrast to Charles Darwin, Wuketits emphasizes that in evolution there is no actual progress because "progress" can hardly be measured and it is difficult to consider one species as the most progressive.[35] Evolution is no rectilinear process but it is irreversible which means that for instance a rhinoceros will always essentially remain a rhinoceros. Wuketits denies an upward development of living species as well as a special place of humanity within nature. He emphasizes however that each living species is unique and therefore that humanity is unique too. Since according to Darwin there are no strict human types but only variations, individuals and populations, the only concrete is always a certain human being of specific looks, distinct ways of thinking, of desires, worries, and hopes. Therefore each of us is "not interchangeable and only identical with itself. Not even identical twins are strictly identical but they differ from each other in certain characteristics."[36]

Since through the history of evolution we all belong together but are unique, no form of discrimination can be justified with the theory of evolution. Wuketits therefore advocates a realistic evolutionary humanism in which we "consider with sympathy other individuals of our species and at least in part other species too" and thereby "make a veritable contribution to a humane

---

[34] Cf. Franz Wuketits, "Zickzackkurs auf dem Grat des Lebens. Zum Weltbild der Evolutionstheorie", in Severin J. Lederhilger, ed., *Den Himmel offen lassen: Der christliche Glaube in der Herausforderung des wissenschaftlichen Weltbildes* (Frankfurt am Main: Peter Lang, 2010), 51.

[35] Cf. Wuketits, "Zickzackkurs", 53.

[36] Wuketits, "Zickzackkurs", 56.

shaping of our future."[37] Wuketits calls himself an atheist but concedes that one can neither prove nor disprove God's existence. Since most people in this world are in a certain way religious, "this concept must be beneficial in an evolutionary way. Evidently we or at least very many of our fellow human beings cannot stand a potential meaninglessness of this world. Therefore humans are inclined to seek answers beyond that which can be experienced."[38] Wuketits considers the faith in God unnecessary for himself but he concedes that other people may hold on to this faith as long as it does not contradict the insights of the natural sciences.

But it is questionable how this concession of a faith in God is more than lip-service when we also notice how Wuketits treats the body-soul or the body-spirit issue. In the light of evolutionary epistemology the dualistic view of body and spirit must be replaced by a monistic one. The spirit is not an independent entity but depends on an active brain. "From an evolutionary point of view the mind is not a category per se nor does it interact with material phenomena (brain) but rather is to be regarded as a *system property* of brain mechanisms; that is, the brain produces mental activities."[39] Mental phenomena depend on the material structures and mechanisms. The spirit is therefore the result of the increasing complexity of the (human) brain and evolution of the spirit is the result of the evolution of the brain. As mentioned previously the spirit is not an (independent) entity but a property of material elements and their interaction. Therefore an evolutionary epistemologist should hold "a monistic view."[40]

From this point of view, life after death is only "an illusion and nothing else."[41] Though Wuketits as a representative of evolutionary teaching lives without such a faith he concedes it to others

---

[37] Wuketits, "Zickzackkurs", 58 f.

[38] Frank Wuketits in Richard Schröder/Franz Wuketits, "Sinn und Unsinn des Glaubens", *Gehirn & Geist* (4/2009), 46.

[39] So Franz Wuketits, *Evolutionary Epistemology and Its Implications for Humankind* (Albany, NY: State University of New York, 1990), 196.

[40] Wuketits, *Evolutionary Epistemology*, 198.

[41] Wuketits, *Evolutionary Epistemology*, 199.

as "a private faith" though one must realize that it does not hold up to scientific scrutiny. According to Wuketits this faith arose from the existential needs, anxieties, hopes, and desires connected with being human. But it has no external anchorage. Once the presumably "hard facts" of the evolutionary theory are isolated from the existential private concerns (of the theologians) there can hardly be a dialogue between the two. This becomes even more evident with the crusader mentality of Richard Dawkins.

The British evolutionary biologist RICHARD DAWKINS (*1941) taught public understanding of science at Oxford University. He considers himself an atheist and humanist and has created his own foundation the *Richard Dawkins Foundation for Reason and Science* which has the aim of furthering scientific education, critical thinking and an understanding of the natural world founded on evidence. Its goal is to eliminate religious fundamentalism, superstition, intolerance, and human suffering. Dawkins supports the *Brights Movement* which was founded in 2003 in the USA but has now spread to many other nations. It wants to advance a public understanding of a naturalistic worldview and further its public acceptance. He is also a supporter of the *British Humanist Association.*

According to Dawkins, the Darwinian theory of evolution has shown through natural selection "living creatures with their spectacular statistical improbability and appearance of design, have evolved by slow, gradual degrees from simple beginnings. We can now safely say that the illusion of design in living creatures is just that – an illusion."[42] Dawkins then concludes: "The factual premise of religion—the God Hypothesis—is untenable. God almost certainly does not exist." The Darwinian doctrine of evolution through natural selection and the understanding of God as a reality exclude each other. Therefore Dawkins goes for the extreme position: "Darwinism impels us to atheism."[43] This

---

[42] Richard Dawkins, *The God Delusion* (Boston: Houghton Mifflin, 2006), 158, for this and the next quotation.

[43] So Alister McGrath, *Dawkins' God: Genes, Memes, and the Meaning of Life* (Oxford: Blackwell, 2005), 51.

means he goes further than Darwin who only admitted that in nature he could only detect a natural causation but no actual plan. Dawkins sees no value in religion as such because it leads the people away from the way things really are. The reason for this is faith, because faith according to Dawkins is "a state of mind that leads people to believe something—it doesn't matter what—in the total absence of supporting evidence. If there were good supporting evidence then faith would be superfluous, for the evidence would compel us to believe it anyway."[44] This means that Dawkins equates faith with credulity.

Once Dawkins has declared God and faith as obsolete he attempts to explain the origin of religion as "a *by-product* of something else."[45] It is not a phenomenon in its own right but an epiphenomenon, because God or some kind of deity does not pertain to any reality. Since Dawkins associates religion with credulity it cannot be anything positive and therefore its survival is likened to a parasite. Just like a parasite in the human mind, religion infests humanity at an early age and as a remnant of one's childhood it remains and then is handed on to the next generation. This view reminds us of SIGMUND FREUD (1856 – 1939) who demanded that humans grow-up and leave behind their infantile and sickly religious ways. For this reason Dawkins compares the concept of God with a meme—a meme is a unit of thought which multiplies through the communication of its carriers—this means a so-called culture gene which in this case is of no value. This meme functions like a virus so that Dawkins can explain the possibility of the multiplying and epidemic character of religion.

The reason why Dawkins is so vehemently opposed to anything associated with religion or with God seems to be in his mindset because, as he says, he is "an intellectual monist."[46] His methodological naturalism has then become exclusive and absolute. It is no longer a heuristic method as employed by religiously unprejudiced scientists but has been changed by him into

---

[44] Richard Dawkins, *The Selfish* Gene (Oxford: Oxford University, 1999), 330.

[45] Dawkins, *The God Delusion*, 172.

[46] Dawkins, *The God Delusion*, 180.

a metaphysical creed. As for instance the Canadian historian of religion Arthur McCalla declares: "Methodological naturalism abstains from making assertions about the nature of reality and instead lays down rules for discovering reliable knowledge about the universe."[47] Dawkins, however, goes beyond these limitations and pursues a metaphysical naturalism which explicitly excludes the existence of non-material forces or figures.

More discerning is EDWARD O. WILSON (*1929) who has been Joseph Pellegrino university research professor in entomology for the Department of Organismic and Evolutionary Biology at Harvard University. Few books have created such a heated debate as did his book *Sociobiology: The New Synthesis* (1975) with which he created a new discipline. Wilson used sociobiology and evolutionary principles to explain the behavior of the social insects and then to understand the social behavior of other living beings, including humans, thus established sociobiology as a new scientific field. He argued that all animal behavior, including that of humans, is the product of heredity, environmental stimuli, and past experiences, and that free will is an illusion.[48]

The grounds for his abolition of free will can easily be gathered from *On Human Nature*. While in *Sociobiology* Wilson paid explicit attention to humanity only in his final chapter, *On Human Nature* (1978), is exclusively devoted to an investigation of human behavior. For Wilson human evolution is determined by genetic evolution. "The genes hold culture on a leash. The leash is very long, but inevitably values will be constrained in accordance with their effects on the human gene pool."[49] In his chapter on religion he shows how this works: "The predisposition to religious belief is the most complex and powerful force in the human

---

[47] Arthur McCalla, *The Creationist Debate: The Encounter between the Bible and the Historical Mind* (Edinburgh: T. & T. Clark, 2006), 193.

[48] For the following cf. Hans Schwarz, "The Interplay between Science and Theology in Uncovering the Matrix of Human Morality," *Zygon* (March 1993), 28: 67–69.

[49] Edward O. Wilson, *On Human Nature* (Cambridge. MA: Harvard University, 1978), 167.

mind and in all probability an ineradicable part of human nature."[50] Since human genes "program the functioning of the nervous, sensory, and hormonal system of the body, and thereby almost certainly influence the learning process", those religious practices that consistently enhance survival and procreation of the practitioners will be favored.[51] Since the actual carriers of biological evolution are not individuals, whether species or single members, but the genes that co-operate in order to survive, traditional religion can be explained "as a wholly material phenomenon."[52] That which looks externally altruistic is genetically egotistic. Natural selection, therefore, does not primarily further the maximizing of personal fitness, but rather the fitness of the whole which is measured in terms of individual success in reproduction plus the reproduction of genetic relatives in which those are preferred that are closer to oneself. While religion itself will endure for long time as a vital force in society, Wilson opts for an alternative mythology by which to explain our special place in the world, "the evolutionary epic."

Edward O. Wilson devoted *Consilience: The Unity of Knowledge* to that alternative mythology. As the word "consilience" or "coherence" says, Wilson wants to achieve a coherence of all knowledge (about the human being), albeit from the angle of genetic evolution. Wilson is convinced that "the conflict between the world views will most likely be settled. The idea of a genetic, evolutionary origin of moral and religious beliefs will be tested by the continuance of biological studies of complex human behavior. To the extent that the sensory and nervous systems appear to have evolved by natural selection or at least some other purely material process, the empiricist interpretation will be supported. It will be further supported by verification of gene-culture coevolution."[53] This means that evolution gained the upper hand. Wilson is convinced that "people need a sacred narrative. ...The true evolu-

---

[50] Wilson, *On Human Nature,* 169.

[51] Wilson, *On Human Nature,* 177.

[52] Wilson, *On Human Nature,* 192, for this and the following quote.

[53] Edward O. Wilson, *Consilience: The Unity of Knowledge* (New York: Alfred A. Knopf, 1998), for this and the following quote.

tionary epic, retold as poetry, is as intrinsically ennobling as any religious epic. Material reality discovered by science already possesses more content and grandeur than all religious cosmologies combined." One can certainly agree with Wilson that "religion must somehow find the way to incorporate the discoveries of science in order to retain credibility. Religion will possess strength to the extent that it codifies and puts into enduring, poetic form the highest values of humanity consistent with empirical knowledge."[54] Yet theologians will have a hard time to accept that religion is void of any trancendent referent. With this premise Wilson is correct that "theology is not likely to surrvive as an independent intellectual discipline."[55] The religion which Wilson still admits is not an indendent phenomenon but contingent upon human genes. We are reminded here of Frank Tipler who wanted to subsume theology and religion under physics, while Wilson opts for biology as the leading category. That with such restrictions religion has no actual future becomes evident in Wilson's more recent book, *The Social Conquest of Earth* (2012).

This book caused a furious review by Richard Dawkins.[56] The reason for this was obvious. Dawkins had felt that Wilson was his ally since both had favored kin selection as the standard process of natural selection. The "selfish gene" helps to propel the individual living being and therewith the species. Now Wilson had changed his mind and claimed that group selection "is responsible for all of our virtues ('honor, virtue and duty'), whole individual selection produces nothing but sin ('selfishness, cowardice and hypocrisy')."[57] But group selection versus the "selfish gene" in kin selection is not the main point in this book but rather the claim that biology has all the answers to the human phenomenon. Wilson is convinced that "the evolutionary model

---

[54] Wilson, *Consilience*, 290.

[55] Wilson, *On Human Nature*, 192.

[56] Cf. Vanessa Thorpe, "Richard Dawkins in Furious Row with E. O. Wilson over Theory of Evolution," *The Observer* (Sunday, June 24, 2012).

[57] So Paul Bloom, "The Original Colonists. *The Social Conquest of Earth* by Edward O. Wilson," *The New York Times* (May 11, 2012), in this very critical review.

within biology ... can explain both the origin and essence of all cultural phenomena" including religion.[58]

Wilson starts with the question: "'*Where do we come from?*' '*What are we?*' '*Where are we going?*'"[59] And he asserts: "Religion will never solve this great riddle. Since Paleolithic times each tribe ... invented its own creation myth." Then he declares:

> The creation myth is a Darwinian device for survival. ... The truth of each myth lived in the heart, not in the rational mind. By itself, mythmaking could never dicover the origin and meaning of humanity. But the reverse order is possible. The discovery of the origin and meaning of humanity might explain the origin and meaning of myths, hence the core of organized religion.

> Could these two worldviews ever be reconciled? The answer, to put the matter honestly and simply, is no. They cannot be reconciled. Their opposition defies the difference between science and religion, between trust in empiricism and belief in the supernatural."[60]

The human condition cannot be solved by religion or introspection but only by scientific investigation. This is the agenda of *The Social Conquest of Earth*. To that end he proposes "that scientific advances, especially those made during the last two decades, are now sufficient to us to address in a coherent manner the questions of where we come from and what we are."[61] Wilson wants to write scientific anthropology in explicit opposition to religion whether Christian or otherwise.

Though Wilson agrees that a clear definition of human nature is an extraordinarily difficult task, he offers such a definition: "Human nature is the inherited regularities of mental development common to our species.They are the 'epigenetic rules', which evolved by the interaction of genetic and cultural evolution

---

[58] So Ted Peters in his extensive and critical review "E.O. Wilson's Conquest of Earth," *Theology and Science* (May 2013), 11:91.

[59] Edward O. Wilson, *The Social Conquest of Earth* (New York: W. W. Norton, 2012), 7, for this and the following quotation.

[60] Wilson, *The Social Conquest*, 8.

[61] Wilson, *The Social Conquest*, 10.

that occurred over a long period in deep prehistory. These rules are the genetic biases in the way our senses perceive the world, the symbolic coding by which we represent the world, the options we automatically open to ourselves, and the responses we find easiest and most rewarding to make."[62] Yet such a definition comes as no surprise. We are in part determined by our genetic endowment and in part by the environment we live in. All of us know this. It is also hard to understand how such truism would mitigate against a theological interpretation, *e. g.*, that God endowed us with certain traits and put us into a certain place to live our lives.

Yet Wilson's "science-based secularism" has no place for God, since God was made in the image of man and not vice versa.[63] Therefore most scientists, Wilson asserts, are not "believers in God." Though most people in America still believe in God and a life after death, this is not longer true for Western Europe. While Wilson has been raised "as a Southern Baptist, an evangelical denomination that includes a large percentage of America's fundamentalist Christians," he tells those who do not want to be ridiculed by reason "think again."[64] Wilson can appreciate the awe and wonder involved in religious rites which serve to unite human tribes as has also been the function of a creation myth. But now "humankind deserves better."[65] Religious faith can now be perceived "as an unseen trap unavoidable during the biological history of our species. And if this is correct, surely there exist ways to find spiritual fulfilment without surrender and enslavement." This estimate comes close to that of Freud who claimed that once religion served a purpose. But as humanity has reached adulthood, it should part from its childhood ways.

Wilson questions "the myths and gods of organized religions … because they encourage ignorance, distract people from recognizing problems of the real world, and often lead them in wrong directions into disastrous actions."[66] Wilson, however,

---

[62] Wilson, *The Social Conquest*, 193.

[63] Cf. Wilson, *The Social Conquest*, 254 f.

[64] Wilson, *The Social Conquest*, 256 f.

[65] Wilson, *The Social Conquest*, 267, for this and the following quotation.

[66] Wilson, *The Social Conquest*, 292.

knows about the real problems, such a habitat destruction, pol-
lution, overpopulation, and overharvesting. Nevertheless, in
conclusion he confesses his

> own blind faith. Earth by the twentysecond century, can be turned, if we
> so wish, into a permanent paradise for human beings, or at least the
> strong beginnings of one. We will do a lot more damage to ourselves and
> to the rest of life along the way, but out of an ethic of simple decency to
> one another, the unrelenting application of reason, and acceptance of
> what we truly are, our dreams will finally come home to stay."[67]

Wilson started this book with emphasis on reason over against
faith and he concludes with a confession of "blind faith." Perhaps
even the most fervent advocate of reason cannot do without a
degree of faith.

On the German side we should mention here the plant physiologist and evolutionary biologist ULRICH KUTSCHERA (*1955)
who teaches at the University of Kassel and who attacks religion
in his own way. He is a member on the governing board of the
Giordano Bruno Foundation and vehemently rejects a creationism which advocates a literal understanding of the first chapter of
the Bible. Moreover he interprets the American distinction between the sciences and the humanities as meaning *Realwissen-
schaften* (i.e., *sciences about reality*) and *Verbalwissenschaften*
(i.e., *sciences of words*). A natural scientist researches things
which really exist whereas a scholar in the humanities is usually
concerned "with weighing the arguments against each other,
interpreting anew, and commenting on that what others have
thought and written about real facts."[68] Biological processes are
foundational for life therefore thinking is also a biological
process. Kutschera claims: "Mental productions can only be
appropriately examined and understood in the light of biology."[69]

---

[67] Wilson, *The Social Conquest*, 297.

[68] So Ulrich Kutschera as quoted in the informative article "Angriff auf
den 'Verbalwissenschaftler'," *Süddeutsche Zeitung* (6.7.2008).

[69] Ulrich Kutschera, "Lobenswerte Bemühungen," *Laborjournal online*
(July 28, 2008).

Problematic in this assertion is again the word "only" because
with this word Kutschera claims exclusiveness for biology as a
scientific discipline. Since he eliminates with one stroke the
Christian faith in creation, the intelligent design theory, and
creationism he does exactly what he forbids the humanities to do,
namely to meddle in the business of another discipline.[70] Kutschera then arrives at the conclusion "that a consistently pursued
(fundamentalist) Christian faith is irreconcilable with evolutionary (naturalistic) thinking."[71] To give his claim more weight
he points out that "the majority of natural scientists are un-
believers." In spite of his (atheistic) engagement he distinguishes
himself from the crusader mentality of Dawkins when he says: "I
have always fully respected the private religious faith of genuine
Christians" because they also acknowledge as consequential his
"atheistic fundamental conviction."[72]

In a research project "Evolution without Genetics – Alternative
Theories in Evolutionary Biology of the 20th Century" supported
by *Deutsche Forschungsgemeinschaft* (*German Research Society*)
he was given the area of creationism as his research project. In the
book resulting from his research Kutschera makes no attempt to
hide his polemic and atheistic attitude. He writes: "The two major
denominations in our country have since then abandoned their
ideological resistance against evolution so that a modern Christian usually today does not see an intellectual opponent in an
evolutionist."[73] It is difficult to recognize what this conclusion

---

[70] Cf. Ulrich Kutschera, *Streitpunkt Evolution: Darwinismus und In-
telligentes Design,* 2nd updated ed. (Münster: Lit, 2007), 92 – 113. In the
chapter "Christlicher Glaube, Darwinismus und Design" (Christian Faith,
Darwinism, and Design) he lumps together his view of the origin of religion, the development of the Christian faith (each time with dates for the
corresponding doctrines), special religious groups, Paley, and Darwin's
reaction to Paley. In conclusion he refutes the faith in a creator. In this
hodgepodge presentation one can hardly detect scientific care.

[71] Kutschera, *Streitpunkt Evolution,* 297.

[72] So Ulrich Kutschera, ed., *Kreationismus in Deutschland. Fakten und
Analysen* (Berlin: Lit, 2007), 362.

[73] Ulrich Kutschera, *Streitpunkt Evolution: Darwinismus und In-
telligentes Design* (Münster: Lit, 2004), 105.

and the preceding condemnation of the churches and of the Christian faith has to do with creationism. In another place he writes: "With Behe (1996) one looks in vain for such Christian religious assertions. On the book cover the claim is made that the author is not a creationist (the biochemist is however a confessing Roman Catholic)."[74] Again we cannot understand what this reference put in parenthesis has to do with creationism unless Roman Catholicism was compared to creationism. According to Kutschera, Siegfried Scherer had claimed in a lecture: "With the origin of life a supernatural intervention is conceivable in the earthly event in analogy into the biblical miracles." In reference to this lecture Kutschera then offers three illustrations: "Biblical miracles in the New Testament: the revivification of the dead young man (below), the multiplication of the bread (center) and the ascension of Jesus (above) (according to a Roman Catholic school Bible, 1930)."[75] If theologians would take illustrations from 1930 in the area of biology to illustrate present-day insights then they would be rightly told that they have misrepresented present-day knowledge. A meaningful dialogue can only be conducted if a dialogue partner takes the other's side seriously. Ulrich Kutschera seems to be far from that.

Scientists not only oppose religion, biblical faith in creation, intelligent design, and creationism. Some of them are among the strongest advocates of these ideas. One example is WERNER GITT (*1937), formerly professor and director of the Department of Information Technology at the Physical Technical Federal Institute in Braunschweig, Germany. He is a Baptist and also a member of the *Study Community Word and Knowledge*. Similar to Kutschera, Gitt is active in the media through many popular scientific books and lectures and also in the internet with MP3 downloads. For an information scientist such as Gitt it is inevitable that next to matter and energy, information is the third foundational entity of all technical and biological processes. This information must have a cause which determines the pertinent code and meaning. This

---

[74] Kutschera, *Streitpunkt Evolution*, 127.
[75] Kutschera, *Streitpunkt Evolution*, 174.

cause Gitt then identifies with God who discloses himself to us in the Bible.[76] This means we have again arrived at an intelligent design analogous to what William Paley advocated.

Yet Gitt goes far beyond Paley because he represents a strict literal biblicism. He writes: "The Bible bears the seal of truth and all its statements are binding, irrespective of whether they are about faith, salvation, or everyday life, or whether they touch on scientific matters."[77] If scientific matters are communicated they naturally go beyond the then available knowledge. They are so to speak anticipatory prophesy. The Bible is then "a book of truth independent of the state of knowledge of that time" even if Gitt finds it helpful if science verifies the Bible.[78] With regard to creation it is evident that the primal atmosphere of the earth was exactly as Genesis 1 described it, meaning "perfectly suited for the animal and plant world and for human beings."[79] Literalistically understood the Bible contradicts present day scientific hypotheses and is true because "the biblical creation narrative is information from God with the seal of absolute truth."[80] With this approach there can be no dialogue with the natural sciences because, just in the opposite way as Tipler has done, Gitt tells the natural sciences what they can regard as truth and what they cannot. The natural sciences are therewith subordinated to biblical literalism.

With his approach Werner Gitt is not a rare exception, since creationism is a worldwide phenomenon and increasing in popularity among the general population. For instance according to a 2006 survey sixty percent of the British population would prefer creationism to be taught in school. In the summer of 2011 in the Netherlands there was the 10th *European Creationist Congress.* According to a 2007 Gallup survey sixty-six percent of the

---

[76] Werner Gitt, *Fragen – die immer wieder gestellt werden* (Bielefeld: CLV, 1990).

[77] Werner Gitt, "The Biblical Teaching Concerning Creation," in Gitt *et al.* (eds.), *Concepts in Creationism* (Herts, Engl.: Evangelical Press, 1986), 20.

[78] Gitt, "The Biblical Teaching Concerning Creation," 26.

[79] Gitt, "The Biblical Teaching Concerning Creation," 32.

[80] Gitt, "The Biblical Teaching Concerning Creation," 36.

American population believes that God created humans approximately 10,000 years ago in their present form. In May 2007 a 70,000 square foot *Creation Museum* opened its gates in Petersburg, Kentucky, near the Greater Cincinnati Airport. The cost of $ 27 million (U.S. dollars) was financed by donations. In its advertisement Johannes Kepler is even cited as a creationist. But where do people get these ideas and from whom is the scholarly acclaim derived?

Two Americans, engineer HENRY M. MORRIS (1918–2006) and Old Testament scholar JOHN C. WHITCOMB (*1943) published *The Genesis Flood* in 1961 in which they claimed to provide a scientific foundation for the late date of creation. They wrote: "A careful study of the Biblical evidence leads us to the conclusion that the Flood may have occurred three to five thousand years before Abraham."[81] This would mean a creation of humanity within the last 10,000 years. In 1972 Morris founded the *Institute for Creation Research* in Dallas, Texas.[82] Soon Morris and his followers attempted to introduce creationism into the curriculum of public schools because they claimed that according to the flood theory the geological formations had been formed in a relatively short time and therefore the theory of evolution was untenable. Since they failed in their endeavors to get creationism into the public schools the *Intelligent Design Movement* was established, since 'design' has been a well-known concept since Paley.

The founder of the *Intelligent Design Movement* is lawyer PHILLIP E. JOHNSON (*1940) and the author of *Darwin on Trial* (Washington, D.C.: Regnery Publishing Co., 1991). He is the cofounder and program director of the *Center for the Renewal of Science and Culture* at the *Discovery Institute* in Seattle, Washington. Against an atheistic theory of evolution he posits a the-

---

[81] John C. Whitcomb/Henry M. Morris, Foreword John C. McCampbell, *The Genesis Flood: The Biblical Record and Its Scientific Implications* (Philadelphia: Presbyterian and Reformed Publishing Co., 1961), 489.

[82] Cf. for the following the well-informed investigation by Ronald L. Numbers, *The Creationists: From Scientific Creationism to Intelligent Design*, exp. ed. (Cambridge, MA: Harvard University, 2006), 312 ff.

istic realism. Another energetic representative of the intelligent design argument is the mathematician and philosopher WILLIAM A. DEMBSKI (*1960) who presently holds a research professorship of philosophy at Southwestern Baptist Theological Seminary in Fort Worth, Texas. – Southwestern Seminary is one of the largest theological educational institutions worldwide. – Dembski is closely connected with the above mentioned *Center.* He asserts that many things in nature are so complex, for instance the molecular machines in a cell, that there is no appropriate natural explanation for them. "The only causal power we know that is able to produce systems like this is intelligence."[83] In advocating an intelligence or an *intelligent design* Dembski does not want to prove the truthfulness of the Christian faith including a creator but he wants to reject the dominant ideology "namely, the scientific and Darwinian materialism that undergirds so many people's rejection of Christianity in Western culture."[84]

The *Intelligent Design Movement* is rejected by many people because it wants to indirectly provide "scientific" verification for the truthfulness of the Christian faith. While the Christian faith rests on certain historical facts, such as the birth and death of Jesus and the assertion of reliable witnesses that Jesus was resurrected, the conclusions from these facts require an existential agreement. Yet with the *intelligent design* argument the conclusions from the historical facts are scientifically "proven." This renders an existential assent unnecessary. One no longer needs faith, because whoever does not agree with the conclusions is just considered to be dumb, or an unbeliever. Siegfried Scherer, who initially was inclined to agree with creationism, also leans in this direction.

SIEGFRIED SCHERER (*1955) is professor of microbiological ecology at the Technical University of Munich in Weihenstephan and there the director of the Central Institute for Food and Research in Nutrition at the same institution. Till 2006 he was

---

[83] William A. Dembski, "Opening Statement," in Robert B. Stewart, ed., *Intelligent Design: William A. Dembski & Michael Ruse in Dialogue* (Minneapolis: Fortress, 2007), 19.

[84] Dembski, "Opening Statement," 22.

honorary president of the study community *Word and Knowledge* and since 2003 he has been a fellow of the *Discovery Institute* in Seattle. In 1998 he and Reinhard Junker of *Word and Knowledge*, had the book *Evolution – A Critical Textbook* (*Evolution. Ein kritisches Lehrbuch*; Gießen: Weyel) published. It was awarded the German prize for textbooks by the association *Learning for the German and European Future* (*Lernen für die deutsche und europäische Zukunft*) which gives awards for books "which inculcate to students reverence for God, love of neighbor, tolerance and ability of dialogue on the foundation of their own ethically strong Christian conviction."[85] He understands evolutionary biology as an important scientific discipline. Yet he emphasizes that empirical science comes to an end when it succumbs to an all-embracing and in part totalitarian naturalism.

According to Scherer intelligent design is not a scientific alternative to biological theories of evolution. "Intelligent design wants to show that the origin of non-reducible complex biological structures cannot be explained by evolution and, among other things, derives from that fact the necessary conclusion of a designer."[86] According to Scherer these gaps of explanation are no proof of God and inexplicability is always a preliminary argument which is tied to our present state of knowledge but is never an ultimate argument. Since every intelligent design movement only assumes that there is a designer behind life, then "this cannot be excluded" with scientific arguments. But this does not mean that the theory of evolution must be abolished. Scherer emphasizes: "Basically I do not know any other scientific alternative to the evolutionary theories in biology." As Scherer notes in the preface to the 6th edition of his textbook on evolution, one must distinguish "between an evolutionism which sets itself as an absolute worldview and the evolutionary theories as scientific

---

[85] http://www.schulbuchpreis.de/kriterien.html. (Stand: August 2011).

[86] Siegfried Scherer, "'Intelligent Design' ist keine naturwissenschaftliche Alternative zu biologischen Evolutionstheorien," in Version 1 of April 20, 2008, source: http://www.siegfriedscherer.de/id.html (date: August 2011), p. 1 for this and the subsequent quotations.

attempts to understand the history of life."[87] While research and interpretation in microevolution is not problematic this is not the case with macroevolution. In macroevolution evolution is interpreted as the leading concept for life as a whole and therefore limits are transgressed because a partial science wants to elevate itself to make assertions about the whole. This recognition of limits being transgressed "is in the same way true for naturalism which equates the scientifically researchable world with reality as well as for the doctrines of creation which correlate with revelation." The possible answers "derived from this contain decisions of faith and determine our view of the world and of humanity; from there far-reaching consequences are drawn for the self-understanding of humanity and its actions." Therefore he understands this book as a contribution to a "scientifically based controversial discourse."

When we compare the position of Scherer with that of Dawkins we notice that no strict scientific decisions have been made but those entailing a worldview. While Dawkins, similar to Gitt, maintains unrestrained a basic decision which he has made, Scherer is much more flexible. He has distanced himself far from creationism though he is still engaged with *Word and Knowledge* whereby it is important for him that he is not be coerced into a peculiar worldview. Scherer is no longer convinced that for all existential questions there is an appropriate theological or scientific answer and he still rejects evolution as the explanation of life in its totality. He emphasizes "that the mechanisms for macroevolution are as unknown as the processes which have led to the origin of life. Whether these are *unsolvable* problems for the teaching of evolution must remain undecided. Undecided issues are an incentive for further research and no necessary argument for a different concept of the origin, such as in the doctrine of creation."[88] Here the scientist Scherer has not yet

---

[87] Reinhard Junker/Siegfried Scherer, *Evolution: Ein kritisches Lehrbuch*, 5 f. (preface), for this and the following quotations.

[88] Scherer, "'Intelligent Design'," 306. Here Scherer shows much more of an affinity to the scientific consensus than in *Die Suche nach Eden. Wege zur alternativen Deutung der menschlichen Frühgeschichte* (Neuhausen-

accepted Martin Luther's insight that God works within and under the forces and powers of this world and therewith also in nature. The rejection of evolutionary theories as a total explanation of life is often understood as a rejection of evolution altogether. Therefore it is understandable that Scherer is confronted with considerable criticism.

7.3 Human Accountability and Religious Naturalism (Hans-Peter Dürr and Ursula Goodenough)

Whether scientists attempt to discern the mind of God, whether they deny that theology can make any credible truth claims, or whether they try to show that the world is indeed God's creation, they show by these very pursuits that scientists are not satisfied with unearthing the factual. They also interpret these facts. Their contradictory claims, however, demonstrate that science is far from being value free. Its application and interpretation has immense implications for our self-understanding and for our own future and the world around us. As Carl Friedrich von Weizsäcker claimed: "Science is responsible for its consequences."[89] This ethical dimension becomes especially noticeable in the work of Hans-Peter Dürr.

HANS-PETER DÜRR (*1929) received his doctorate under the guidance of the Hungarian-American nuclear physicist EDWARD TELLER (1908–2003), generally known as "the father of the hydrogen bomb." Then he worked under Werner Heisenberg, became director of the Max Planck Institute for Physics in Munich and was awarded the alternative Nobel Prize in 1987. Dürr is a member of the *Club of Rome* and in addition to his scientific research is concerned with issues of the future of the world be-

---

Stuttgart: Hänssler, 1991), 179, a publication he edited. There he still wrote: "The ages of the patriarchs provided in Genesis, the size of the initial population, and the available time frame for the increase in population are not untenable biologically speaking."

[89] Carl Friedrich von Weizsäcker, *Wahrnehmung der Neuzeit* (Munich: Carl Hanser, 1983), 428.

cause, as he says, the world can live without humans but humans cannot live without the world. For him scientific knowledge has no claim of exclusivity. "Science tells us what is, but it does not give us any answer what should be, how we should act. In order to be able to act humans need insight which goes beyond scientific knowledge, they need guidance through something transcendent."[90] Yet the scientific way of seeing things and the rapid technological progress block the view to the transcendent and its necessity for life. Only a few people understand the consequences of modern physics because it is highly abstract. Therefore most people remain caught-up in the classical mechanistic and deterministic worldview of the 19th century. But as we have heard from him: "Physics and transcendence in the conception of present-day physicists are no longer related to each other in an antagonistic way but rather in a complementary one," even if this complementarity is interpreted by theologians and scientists in different ways.[91]

For Dürr it is a matter of fact that scientists cannot exclusively devote themselves to the knowledge of relationships in nature. Because of the public political expectation that their knowledge can be applied they must also assume responsibility in society. Science has created tools which in principle can lead to the annihilation of humanity and its environment. Responsibility does not mean for Dürr that scientists "should take the direction of nature consciously in their hands and do this with utmost sensitivity and caution."[92] For Dürr this means however responsibility to safeguard peace, especially in view of the nuclear threat, but also to supply energy without the problematic nuclear energy. Ecological problems which have been caused by our intervention

---

[90] Hans-Peter Dürr, *Physik und Transzendenz: Die großen Physiker unseres Jahrhunderts über ihre Begegnung mit dem Wunderbaren*, 8, in his preface.

[91] Dürr, *Physik*, 11.

[92] Hans-Peter Dürr, *Das Netz des Physikers: Naturwissenschaftliche Erkenntnis in der Verantwortung* (Munich: Carl Hanser, 1988), 11, for this and the following quotations. In this publication Dürr clearly shows both the responsibility and the problems.

in nature also require the engagement of scientists because only through the application of scientific knowledge could we drastically delve into the web of nature.

It is important for Dürr that we become more cautious in the interdisciplinary dialogue because "the insight in our limitations leads to an appropriate degree of modesty on which we can erect a differentiated worldview. We must develop the ability to understand each phenomenon as part of a greater wholeness. This wholeness can be experienced not by the dissecting method of science but by the subjective inner view, through a religious perception."[93] Dürr also seeks out the dialogue with other sciences and with theology to attain an understanding of reality characterized by wholeness. This is also shown in a more recent publication *Love – Primal Source of the Cosmos: A Dialogue in Science and Religion* (*Liebe – Urquelle des Kosmos: Ein Gespräch über Naturwissenschaft und Religion*; 2008) which resulted from a conversation between Dürr and RAIMON PANIKKAR (1918–2010). There he talks with an Indian philosopher of religion about a variety of topics such as "dying, death, and life beyond" and "modern science and knowledge." Dürr is "convinced of an open world of potentiality because the statistically, the solid, and the determined is now in need of being explained."[94] The objective outside world is only an 'as-if' construction for humanity serving its life. Besides this, according to Dürr, there is also "an inner view" in which I am inseparably connected with everything. Religion demands this "inner view" whereas the sciences limit themselves to formulating experiential knowledge which is indispensable for our life. With this dual view of the world theology and the sciences can enter into a mutually enriching dialogue without needing to curtail their own competencies.

---

[93] So Hans-Peter Dürr in his preface to an interdisciplinary dialogue, *Gott, der Mensch und die Wissenschaft* (Augsburg: Pattloch, 1997), 8 f.

[94] Hans-Peter Dürr/Raimon Panikkar, *Liebe – Urquelle des Kosmos: Ein Gespräch über Naturwissenschaft und Religion,* Roland R. Ropers, ed. (Freiburg: Herder, 2008), 37 and 40.

Coming from a different tradition as does Hans-Peter Dürr, but akin to him in her concern for creation is Ursula Goodenough (*1943). She is Professor of Biology at Washington University in St. Louis, a frequent contributor to *Zygon,* and much indebted to *IRAS* in her thoughts about the relationship between religion and science.[95] She came to be especially known for her 1998 book *The Sacred Depth of Nature.* The agenda for this book, she writes, "is to outline the foundations for a planetary ethic."[96] This ethic must be based on religion, because "without a common religious orientation, we basically don't know where to begin, nor do we know what to say or how to listen, nor are we motivated to respond."

But how do we arrive at a common religious persuasion? Goodenough answers that we must start with how things are, the scientific account of Nature. – It is telling that Nature is capitalized here. – This account which she calls the "Epic of Evolution" will make us feel religious. She appeals here to the beauty of nature but not to its workings, well-knowing that its workings can be very cruel as Darwin had frequently emphasized. Considering the grandeur and beauty of nature will fill us with joy and thanksgiving. From natural reality religious emotions can be elicited and therefore Goodenough is convinced that "the story of Nature has the potential to serve as the cosmos for the global ethos that we need to articulate."[97] On the premise that nature is sacred Goodenough advocates a religious naturalism which emphasizes the significance of religion without adhering to any kind of theism. As Willem Drees emphasizes: "Religious naturalism also takes shape through stories. The evolutionary epic serves as a master narrative, but there are smaller stories that evoke attitudes and feelings alongside philosophical essay that convey intellectual claims. Ursula Goodenough's *The Sacred Depth of N*ature is an example."[98]

---

[95] So Ursula Goodenough, *The Sacred Depths of Nature* (Oxford: University Press, 1998), xi, in her personal reflections.

[96] Ursula Goodenough, *The Sacred Depths of Nature,* xvf., for this and the following quote.

[97] Ursula Goodenough, *The Sacred Depths of Nature,* xvii.

[98] Willem B. Drees, "Religious Naturalism and Science", in *The Oxford*

Goodenough wants to anchor a global ethic both in an understanding of human nature and in understanding the rest of nature. While this starting point certainly deserves merit the questions is how to proceed from the given, that is humanity and nature, to a certain direction in which to move. Perhaps she is aware of this problem, since she admits: "I will not presume to suggest what this ethos might look like. Its articulation must be a global project."[99] But she is convinced that "the existence of all this complexity and awareness and intent and beauty, and my ability to apprehend it, serves as the ultimate meaning and the ultimate value. The continuation of life reaches around, grabs its own tail, and forms a sacred circle that requires no further justification, no Creator, no superordinate meaning of meaning, no purpose other than that the continuation continue until the sun collapses or the final meteor collides. I confess a credo of continuation. And in so doing, I confess as well a credo of human continuation".[100] Through her appearance at PBS, History Channel, and National Public Radio and her insistence on the sacredness of nature and the awe for the "Epic of Evolution" she has exerted considerable influence and disseminated her religious naturalism which needs no personal God since God and nature blend together.[101]

Handbook of Religion and Science, Philip Clayton, ed., Zachary Simpson, assoc. ed. (Oxford: University Press, 2006), 120.

[99] Ursula Goodenough, *The Sacred Depths of Nature*, xvii.

[100] Ursula Goodenough, , *The Sacred Depths of Nature*, 172.

[101] For further information on religious naturalism cf. Michael Hogue, *The Promise of Religious Naturalism* (Lanham, MD: Rowman & Littlefield, 2010) and Loyal Rue, *Nature Is Enough: Religious Naturalism and the Meaning of Life* (New York: Suny Press, 2011).

8. Partners from Theology

Turning to the theologians as dialogue partners in the dialogue between theology and the natural sciences we quickly notice that in the Anglo-Saxon world there are considerably more theologians dealing with issues of the natural sciences than in the continental European world. The reason for this is the fact that as a result of the Enlightenment natural theology was always held in high esteem in the Anglo-Saxon world and secondly the neo-Reformation theology of Karl Barth had never been accepted there on a large scale. His theology was considered to be more evangelical and his person of course was always associated with his fierce resistance against the Hitler regime. Moreover, professorships in German language universities are more traditional in focus than in Anglo-Saxon countries where professorships are reoriented much more quickly toward present day issues.

8.1 Different Traditions (Russell, Hefner, Polkinghorne, McGrath, Deane-Drummond, Ijjas, Drees)

Those who have spent or are spending most of their energy in the dialogue with the natural sciences come from very different traditions both in terms of their denominational and national background as well as their theological persuasion. While Robert Russell (United Church of Christ), John Polkinghorne and Alister McGrath (Anglicans), Celia Deane-Drummond (Roman Catholic) and Anna Ijjas present a more conservative spectrum, Willem

Drees comes from a liberal tradition and Philip Hefner (Lutheran) from a more mainline perspective. Yet they all have made an important impact on the dialogue.

We already mentioned Robert John Russell several times as a representative of the present generation. He is Ian G. Barbour professor for theology and the natural sciences at the *Graduate Theological Union* in Berkeley and at the same time director of *CTNS*, a position which he largely financed himself. Since Russell received his doctorate in physics on the same day in 1976 as he was ordained pastor of the United Church of Christ shows that with him the natural sciences and theological reflection belong together. He began his teaching career at Carleton College in Northfield, Minnesota where Barbour also taught. Through the initiative of Barbour and Russell and of two members of the faculty of the Graduate Theological Union the foundation of *CTNS* was established and in 1981 Russell's work began there.[1]

The understanding of the relationship between theology and the sciences at that time was not coined by the image of animosity but by the image of two different languages, that of facts and that of values. This means the two realms were separated and were unable to find a mutually acceptable place in the then prevailing culture where they could meet. Russell, however, wanted to move beyond this co-existence to a dialogue and to a mutually enriching interaction.

There are especially four areas in which he has been active since then:

1. The Big Bang and the finitude of time. He asserts that "the theological meaning of finitude is that the universe is dependent on God for its existence, whatever its duration, and he

---

[1] For this and the following information see Ted Peters, "Russell's Contribution to the Theology and Science Dialogue", in *God's Action in Nature's World. Essays in Honor of Robert John Russell,* Ted Peters/Nathan Hallanger, eds. (Aldershot, Hampshire: Ashgate, 2006), 6.

cautions against identifying the idea of creation too closely with the finitude of past time."[2]

2. On the quantum level there is no divine activity intervening from above. Russell refers here to the Heisenberg Uncertainty Principle and emphasizes that this uncertainty does not mean a limitation of present day scientific knowledge. God, however, can realize a multitude of possibilities without circumventing the laws of quantum physics. Though one could expect that the individual quantum events are averaged out through the statistics valid for large groups, as one can see for instance in the biological process where a single event can have far-reaching consequences. "A single photon disrupting a hydrogen bond in a strand of DNA might alter evolutionary history."[3]

3. Order, disorder and the problem of evil. Here Russell is concerned with the question of theodicy which pertains to the question of why God, who is good, allows so much evil to take place in nature and in human life. He shows parallels between the larger entropy by which the disorder in the physical world can be measured and the presence of suffering and evil. Similarly to physical disorder, Russell suggests, suffering and evil could contribute to new creative possibilities. Ultimately they will be transformed in a future life beyond death.

4. Finally he is concerned with the future of the cosmos and eschatology. He shows that from a scientific viewpoint the world has no future because either expansion continuously increases so that the universe becomes too cold and life can no longer be sustained or conversely that there will be a contraction through which the universe becomes too hot and all life will be ultimately extinguished. Either prospect makes life on earth transient and meaningless. The resurrection of Christ, however, "has given the Christian community grounds for hope in God's transformative power in a cosmic 'New Creation' as well as in individual life beyond death."

---

[2] So Ian Barbour in his preface to Robert John Russell, *Cosmology: From Alpha to Omega. The Creative Mutual Interaction of Theology and Science* (Minneapolis: Fortress, 2008), iv.

[3] Barbour in his preface, v, for this and the following quotation.

5. In his book *Cosmology: From Alpha to Omega* Russell brings together his various contributions such as creation out of nothingness in the light of scientific cosmology, non-interventionist divine action, the problem of natural evil in physics and biology, and also the future of the cosmos and eschatology. While he has published numerous essays and edited a number of books resulting from conferences, with regard to a book of his own on the subject he has been rather reticent.

Even his most recent publication, *Time in Eternity: Pannenberg, Physics, and Eschatology in Creative Mutual Interaction* (University of Notre Dame, 2012), he not only dedicated this volume to Pannenberg, but engages in conversation with him throughout the book, since he finds Pannenberg's insight on time and eternity most fruitful in the dialog with science. Yet, as is typical for Russell, he does not just restate what others have said. Regarding Pannenberg he says: "To deepen that engagement on the theme of 'time and eternity,' much of his work used in this volume must first be reconstructed within the framework of contemporary physics, cosmology, and mathematics."[4] Russell is convinced that "Pannenberg's reconstructed theology might illuminate the search for criteria of choice among existing scientific theories as well as the search to construct new scientific theories."[5] He concludes from Pannenberg that "the distinction between events in time will be sustained in eternity while the separation between events in time will be overcome in eternity."[6] Eternity is neither conceived of as a conflation of all events nor a separation of every event from all others. Of course, this has implications of how science views time.

This means that in the dialogue between theology and the natural sciences Russell wants to attain more than theology which simply interprets scientific facts theologically. He presses for an actual creative mutual interaction between the two disciplines. For

---

[4] Robert Russell, *Time in Eternity: Pannenberg, Physics, and Eschatology in Creative Mutual Interaction* (Notre Dame, IN: University of Notre Dame, 2012), 9 f.

[5] Russell, *Time in Eternity*, 317.

[6] Russell, *Time in Eternity*, 319.

instance theology could pose questions to the scientists and suggest topics and conceptualities which are helpful for their own scientific criteria and research. Therewith "theology can indirectly influence science as a whole as well as to inspire specific research programs in science. In short, each side can find new insights and challenges from the other while retaining their independent entities as authentic fields of discourse and discovery."[7] This conversation or even collaboration between theology and the sciences could for instance be expanded to the relationship between Christian eschatology and a cosmic future. Russell writes:

> If the processes of nature which science describes are due ultimately to God's ongoing action as Creator, and since God is free to act in radically new ways in history and in nature, then the future, shaped by God's new way of acting as begun on Easter, will not be what science predicts based on the present creation. Instead it will be based on a radically new kind of divine action which began with the resurrection of Jesus at Easter and which inaugurated the transformation of the universe into the eschatological New Creation.[8]

Russell's theologically conservative approach naturally contains a great challenge for the sciences because they would have to acknowledge that theology understands God as the creator, sustainer, and completer of the world and that it interprets the possibility given by science in that direction.

In his guidelines for constructive work in science Russell therefore suggests to search for "a richer theological conception of nature both as creation *and* as a new creation [which] can generate important revisions in the *philosophy of nature* that currently underlies the natural sciences, the philosophy of space, time, matter, and causality in contemporary physics and cosmology."[9] But theology too is not unaffected by this new orientation. For instance the understanding of time in physics and cosmology would have to be carefully discussed and the theological topics in relation to this present understanding of time

---

[7] Russell, *Cosmology*, 22.

[8] Russell, *Cosmology*, 24.

[9] Russell, *Cosmology*, 311.

would have to be reformulated. The theological categories of eternity and omnipresence must be interpreted for instance in the light of the special and the general theory of relativity in view of the presently flowing time and time as duration. "Similar theological reconstructions will hold for the treatment of time and space in quantum mechanics."[10] Of course such a dialogue presupposes considerable knowledge of the other respective side, a precondition for researchers who make this dialogue their special area of expertise.

There are also theologians who did not study one of the sciences in their academic education but nevertheless engaged intensively with issues in the sciences. Among these can be counted the Lutheran systematician PHILIP HEFNER. Though he wrote numerous editorials for the journal *Zygon* and edited several books, entire books by him are rare. Yet with the publication of one book *The Human Factor: Evolution, Culture and Religion,* essentially a collection of essays, he attracted considerable attention especially when he named the human being *created co-creator.* This designation was picked-up by many others. Hefner points out that "the model of biocultural evolution requires this concept of genes and culture as two streams of information that comprise the human being."[11] While the genes make available in the human body the genetic information for the human constitution culture gives the human being information from outside the body for the cultural formation. Both come together in the central nervous system to form a human being as a biological cultural being. Hefner further asserts that "the earthly career of this two-natured creature, the human being, is characterized through and through by the marks of being conditioned and also of being free."[12] This determination has its roots in the evolutionary development because the human being originates from a deterministic process which dates back to the origins of the universe. Within this deterministic

---

[10] Russell, *Cosmology,* 314.

[11] Philip Hefner, *The Human Factor: Evolution, Culture and Religion* (Minneapolis: Fortress, 1993), 29.

[12] Hefner, *The Human Factor,* 30.

evolutionary process, however, freedom has developed which has its roots in a genetically controlled adjusted plasticity of the human phenotype.

This freedom shows itself according to Hefner in three ways:

1. In the exploration of the environment to conduct itself in it appropriately.
2. In the self-conscious deliberations of alternative decisions and ways of conduct.
3. In a supportive societal network that allows for explorations by individuals and at the same time demands that group relationships and the welfare of others and also of society as a whole be respected.

This last mentioned characteristic is the biological reason for values and that which can largely be designated as morals. The human being interacts with the environment and with other surrounding human beings in such a way that the well-being of the individuals and of society is reached. These three elements largely coincide with an empirical description of humanity. Then, however, Hefner adds a theological interpretation which he describes as follows:

1. The human being is created by God to be a co-creator in creation that God has brought into being and for which God has purposes.
2. The conditioning matrix that has produced the human being – the evolutionary process – is God's process of bringing into being a creature who represents the creation's zone of a new stage of freedom and who therefore is crucial for the emergence of a free creation.
3. The freedom that marks the created co-creator and its culture is an instrumentality of God for enabling the creation (consisting of the evolutionary past of genetic and cultural inheritance as well as the contemporary ecosystem) to participate in the intentional fulfillment of God's purposes.[13]

---

[13] Hefner, *The Human Factor*, 32.

This in a nutshell is the thesis of Hefner's endeavor to understand the evolutionary process as a God-caused process which brought forth human beings and in which, so to speak, as God's representatives the human beings advance their own being and the surrounding world in the light of God's intentions.

The decisive question is how the human being knows what God's intentions are. Here Hefner writes: "Nature is the medium through which the world, including human beings, receives knowledge, as well as grace. If God is brought into the discussion, then nature is the medium of divine knowledge and grace."[14] This would mean that a self-disclosure of God does not bring anything additional to human insights of what humans cannot already know by themselves from the world. The world, or rather nature, becomes a means of revelation. Hefner seems to consider nature as a means of God's grace so that grace works through nature.

So as not to neglect the spiritual side completely, Hefner defends against process theologians the idea of creation out of nothingness. He contends: "The creation out of nothing puts the reality of goodness, in conjunction with the way how things really are and the intimate relationship between these and the created order, at the very origin of nature."[15] Creation out of nothingness is therefore an important basis on which one must interpret nature as God's good creation. It says "that God is the sole source of all that is, and that God has created freely and without coercion or limitation."[16] Since according to Luther the finite can encompass the infinite, as can be seen with the human being Jesus of Nazareth who is at the same time divine, grace preserves nature, but does not destroy it. It works in it and "leads nature to its fulfillment."[17]

God can work in and through nature and a human being can be understood as being created in God's image. But the latter should not be misunderstood anthropocentrically but as saying that a

---

[14] Hefner, *The Human Factor*, 42.

[15] Hefner, *The Human Factor*, 231.

[16] So Philip Hefner, "Biocultural Evolution and the Created Co-Creator," in *Science and Theology. The New Consonance*, ed. Ted Peters (Boulder, CO: Westview, 1998), 183.

[17] So Hefner, *The Human Factor*, 234.

human being is the free creator of giving meaning and who is active endowing things with meaning and therefore is responsible for these activities as well as for the underlying endowment of meaning.[18] Hefner sees here for humanity great freedom and not only the obligation to administer creation in the line of God's intention.

Since humans are God's co-creators one can also understand conversely God as a "participant" in the (technological) achievements of humanity.[19] God does not tell us how we should exactly conduct ourselves but we are free to shape that which has not been realized. We are the ones who create and thereby discover what is important for us for life. We have been created with an innate obligation to self-transcendence to contribute to the further evolution of the created world. In order that this command of being in-charge of creation does not run aground, and Hefner sees threatening signs for that, he directs our view to Jesus as a paradigm of what it means to be human and to live according to God's image.

Firstly Hefner deals here with Jesus' self-giving love which can be interpreted with the metaphor of sacrifice in evolutionary terms, as altruistic in relation to that which really is.[20] Another important symbol for Hefner is Christ as the second Adam because through it he can show the picture and the attainment of that which the first Adam can become and that for which he was intended. Through this the possibility is opened for us to realize what Jesus has shown us as God's aim. Hefner therefore calls for a "revitalization of our mythic and ritual systems, in tandem with scientific understandings, so as to reorganize the necessary information. This may help us to put our world together, to discern *how things really are,* so that we can test which actual everyday praxis is most adequate for us."[21] Theology and the natural sciences cooperate, according to Hefner, to determine the location of humanity in the world and humanity's conduct in it. Yet we can easily detect that in this interplay, even when Hefner argues as a

---

[18] Cf. Hefner, *The Human Factor*, 239.

[19] Cf. for the following Philip Hefner, *Technology and Human Becoming* (Minneapolis: Fortress, 2003), 79 f.

[20] Cf. For this and the following Hefner, *The Human Factor*, 248.

[21] Hefner, *The Human Factor*, 278.

theologian, the natural sciences retain the upper hand and theologians must ask themselves how they can contribute their foundational theological insights in view of the scientific facts.

There are many other theologians in the USA who engage in the dialogue with the sciences. There is for instance NANCEY MURPHY (*1951). She received her BA from Creighton University majoring in philosophy and psychology. Then she obtained a PhD in philosophy of science from the University of California in Berkeley and finally in 1987 received her ThD from the *Graduate Theological Union*. She is currently professor of Christian philosophy at Fuller Theological Seminary at Pasadena, California, and a prolific writer. Her first book, *Theology in the Age of Scientific Reasoning* (Cornell 1990) won the Award for Excellence of the American Academy of Religion. She also serves on the board of directors of *CTNS* and is heavily involved in the research enterprises conducted there. Next to her academic acclaim she is an ordained minister in the Church of the Brethren. In her work she relies especially on the philosopher of science IMRE LAKATOS (*1922) who claims that in science so-called research programs, and not individual theories, compete with each other. Each step of a research program which has hypotheses should increase the content of the program. The rejection of such a program is only possible if a better theory is available that anticipates new facts. TED PETERS (*1941), a Lutheran systematic theologian who has collaborated for many years with Robert Russell and has especially been interested in the ethical consequences of the genome project, has also dealt largely with the dialogue between theology and the sciences. He is strongly influenced by the eschatologically oriented approach of Wolfhart Pannenberg. Finally we should mention the Lutheran theologian PAUL SANTMIRE (*1935) who has dealt primarily with ecological topics and has objected to the charge that the Christian faith is anthropocentric and therefore destructive of the environment. His book *Brother Earth: Nature, God, and Ecology in a Time of Crisis* which was published in 1970 had a considerable influence of raising ecological awareness, as did his 1985 publication *The Travail of Nature: The Ambiguous Ecological Promise of Christian Theology*. Altogether we notice that many theologians in the USA seek and advance the dialogue between theology and the natural sciences. But in the so-called Old World too this dialogue has attained new significance. From Great Britain we must especially mention John Polkinghorne.

JOHN POLKINGHORNE (*1930), a former professor of mathematics and physics at the University of Cambridge in England was later on ordained as Anglican priest and has become one of the most prominent representatives in the dialogue between theology and the natural sciences. To believe in God in an age of science means for him to have the assurance "that there is a Mind and a Purpose behind the history of the universe and that the One whose veiled presence is intimated in this way is worthy of worship and the ground of hope."[22] Therefore it is not surprising that according to Polkinghorne "science and theology have a fraternal relationship and they are complementary, rather than antithetic, disciplines. Yet each surveys the one world of experience from its own perspective and therefore there are possible points of contact, or even conflict, between them."[23] He thinks that "there is an inescapable interaction between science and theology, as the whole intellectual history from Copernicus through Darwin to the present day makes abundantly clear."[24] This history is by no means one of a continuous battle as it has often been depicted by historians and Polkinghorne even thinks that "the two disciplines need each other."[25] For once if theology takes seriously its own assertions that the world is God's creation then it must be open to learn from the natural sciences what the world actually is. Conversely if the natural sciences refuse to dialogue with theology then they do not permit an explanation of the world which goes deeper than that which the natural sciences

---

[22] John Polkinghorne, *Belief in God in an Age of Science: The Terry Lectures* (New Haven: Yale University, 1998), 1.

[23] John Polkinghorne, *Science and Creation: The Search for Understanding* (Boston: Shambhala, 1989), xii.

[24] John Polkinghorne, "Creation and the Structure of the Physical World," *Theology Today* (1987/88), 44/1:67.

[25] Polkinghorne, "Creation and the Structure," 68.

can adduce already by themselves. Therefore Polkinghorne concludes: "Religion without science is confined; it fails to be completely open to reality. Science without religion is incomplete; it fails to attain the deepest possible understanding."[26]

Polkinghorne does not want to say that the natural sciences can only go to a certain point and then that theology must assume the explanation of reality. Any question which is posed by the natural sciences can also in principle be answered by the natural sciences. In this respect the natural sciences do not need any help from theology. If one would insist on that then one would return again to the God of the gaps. In a similar way theology is concerned to fathom its own phenomena. Here the natural sciences can by no means deny or approve the claims of theology. If it were otherwise then we would return again to the scientific superstition of the 19th century in which the natural sciences defined the space in which theology could still move. Nowadays the natural sciences as well as theology have their own respective autonomy.

But theology and the natural sciences relate themselves to the nearly identical realm namely the world and everything in it. In researching this arena "sciences is broadly concerned with process, with asking the question how things happen" while theology "is broadly concerned with meaning and purpose, with asking the question why things happen."[27] Since they pose these two different questions which yet in some sense belong together theology can provide answers for the natural sciences "to those meta-questions which arise from science but which are not themselves scientific in character."[28] This means theology deepens the results of the natural sciences to quench ultimately the thirst for understanding which the natural sciences as a limited means of obtaining knowledge with a limited realm of research cannot accomplish by themselves. On the other side the natural

---

[26] Polkinghorne, *Science and Creation*, 97.

[27] John Polkinghorne, "Reckonings in Science and Religion," *Anglican Theological Review* (1992), 74/3:376.

[28] John Polkinghorne, *Reason and Reality: The Relationship between Science and Theology* (Philadelphia: Trinity Press, 1991), 75, for this and the following quote.

sciences can tell theology "what the physical world is actually like." If theology wants to advance the doctrine of creation then it must take into account the history of the universe in all its intricate detail which can best be adduced by the natural sciences.

Polkinghorne summons theologians and natural scientists to consider their own tasks and the necessity of entering into dialogue with each other. In so doing there are not only agreements, but as Polkinghorne himself admits also disagreements, for instance in eschatology. According to cosmologists, who concern themselves with the future,

> ultimately it will all end badly, either in the universe's decay or its collapse. ... Theology must take seriously this prognostication of eventual futility. Christian thought, however, has never subscribed to a merely evolutionary optimism. Only God himself can be the ground of a true and everlasting hope. If there is destiny beyond death for ourselves, both must depend upon some great redemptive act of God, triumphing over futility.[29]

This means that ultimately theology cannot agree with the natural sciences in its predictions concerning the exploration of ever greater details of our space-time continuum. Theology must reject finitude the only thing which can be predicted by the natural sciences. There is no ultimate future for anything within our universe if one does not take God into consideration. True future and true fulfillment for the individual and for the totality can only result from God's grace and God's own activity.

Polkinghorne knows that the best dialogue occurs between a competent scientist who does not presume to be an amateur theologian and a competent theologian who does not pretend to be an amateur scientist. Both disciplines must be appropriately distinguished from each other but also related to each other. If the two disciplines are totally separate then they will pass by without noticing each other. If, however, they approach each other but do not appropriately distinguish their own realms of expertise from each other then there is the strange encounter between a theo-

---

[29] Polkinghorne, "Reckonings in Science and Religion," 379 f.

logical natural science and a scientific theology whereby both forget their actual competency and their respective area of research. In making available its own insights natural science can show theology that one can talk about God's activity, his creation, providence, and even about eschatology in a scientifically competent way whereas theology shows that a natural science which teaches nature without reference to God as creator knows no compassion nor that it can be directive.

WHILE Polkinghorne arrived at theology after a distinguished career in the sciences, ALISTER EDGAR MCGRATH (*1953) moved immediately from his education in the natural sciences to theology. He is an Irish systematic theologian and currently professor of theology, ministry, and education at Kings College London and head of the Centre for Theology, Religion and Culture. Previously he taught both at the University of Oxford and at Cambridge University. McGrath holds two doctorates from the University of Oxford, one in molecular biophysics and another one in theology. He is an Anglican and is ordained within the Church of England. He is a prolific writer in many different areas, but the interaction of Christian theology and the natural sciences has been a major theme of his research work, and is best seen in the three volumes of his *Scientific Theology* (2001–3). He attempted to explore there the methodological parallels between theology and the natural sciences "to develop and extend Torrance's vision of theological science."[30] In recent years, he has been especially interested in the emergence of "scientific atheism", and has researched the distinctive approach to atheist apologetics found in the writings of Richard Dawkins.

As he tells in his book *The Foundations of Dialogue in Science and Religion* he was asked already in 1978 by Oxford University Press to write a book responding to Richard Dawkins *The Selfish Gene*.[31] Yet only after extensive studies in theology did he write twenty years later a book in which he attempted "to establish the

---

[30] Alister McGrath, *A Scientific Theology*, vol. 1: *Nature* (Edinburgh: T & T Clark, 2001), 76.

[31] Cf. Alister McGrath, *The Foundations of Dialogue in Science and Religion* (Oxford: Blackwell, 1998), 6.

foundations for dialogue in science and religion by exploring the critically important area of methodology."[32] In four areas he lines out and compares the methods in theology and science, namely in the quest for order, in the investigation of the world, in dealing with the reality of the world, and finally in how they represent the world. He concludes his investigation with a reference to John Calvin. For Calvin "the natural order is a theatre in which the glory of God is displayed to humanity, and through which something of the majesty of God can be known."[33] Therefore it is not surprising that McGrath's main research interest is the area of thought traditionally known as "natural theology", which is experiencing significant renewal and revitalization at the moment.

McGrath addressed this theme in detail at his 2008 Riddell Memorial Lectures at Newcastle University, England, where he stated: "If the heavens really are 'telling the glory of God' (Ps. 19:1), this implies that something of God can be known through them, that the natural order is capable of disclosing something of the divine. But it does not automatically follow from this that *human beings*, situated as we are within nature, are capable unaided, or indeed capable under any conditions, of perceiving the divine through the natural order."[34] McGrath opposes the idea that natural theology "designates the enterprise of arguing directly from the observation of nature to demonstrate the existence of God."[35] As the approaches of Polkinghorne and Dawkins show, nature is open to multiple interpretations. Yet McGrath is convinced that "nature reinforces an existing belief in God through the resonance between observation and theory."[36] Moreover, natural theology "enables a deepened appreciation of nature." To facilitate the dialogue between theology and the natural sciences McGrath wrote a primer on that dialogue, a book in which he explores and explains "the main themes and issues in

---

[32] McGrath, *The Foundations of Dialogue,* 29.

[33] McGrath, *The Foundations of Dialogue*, 208.

[34] Alister McGrath, *The Open Secret: A New Vision for Natural Theology* (Oxford: Blackwell, 2008), 1 f.

[35] McGrath, *The Open Secret*, 4.

[36] McGrath, *The Open Secret*, 18, for this and the following quotation.

the study of religion and the natural sciences," and also introduces the major voices in this dialogue.[37]

In the 2009 Hulsean Lectures at the University of Cambridge
McGrath provides a largely historic narrative of the interaction
between Darwin's notion of evolution and natural theology.
Having laid out the breadth of the terms "Natural theology" and
"Darwinism", McGrath provides a historical exposition of Darwin's ideas and claims and the British natural theology tradition
beginning with the emergence of that tradition at the end of the
17th century till the eve of the Darwinian revolution in 1860. He
shows that some traditional Christian thinkers saw Darwin's new
theory "as a threat to the way in which they had interpreted their
faith. Yet here can be no doubt – for the historical evidence is
equally clear – that other Christians saw Darwin's theory as offering new ways of understanding and parsing traditional
Christian ideas."[38] McGrath points out that "Darwin's theory
appears to have met more sustained opposition from the scientific community than from its religious counterpart."[39] He reminds us that Darwin was neither an atheist as some still portray
him, nor did he eliminate teleology from nature. McGrath refers
here to Thomas H. Huxley (1825 – 95), the staunch advocate of
Darwin's theory of evolution, who stated that the "teleological
and the mechanical views of nature, are not, necessarily, mutually
exclusive."[40] Teleology was simply modified to cope with the
theory of natural selection. McGrath is confident that "the enterprise of natural theology has, if anything, been given a new
lease on life through the rise of evolutionary thought, partly by
being liberated from the intellectual and spiritual straitjacket
within which Paley's approach has unhelpfully confined it."[41] The
point is no longer to demonstrate that God is indeed the creator of

---

[37] Alister McGrath, *Science & Religion: A New Introduction*, 2nd ed.
(Oxford: Wiley-Blackwell, 2010 [2009]), vii.

[38] Alister McGrath, *Darwinism and the Divine: Evolutionary Thought
and Natural Theology* (Oxford: Wiley-Blackwell, 2011), 150.

[39] McGrath, *Darwinism and the Divine*, 151.

[40] McGrath, *Darwinism and the Divine*, 164, where he quotes Huxley.

[41] McGrath, *Darwinism and the Divine*, 280.

the world. Starting, as McGrath does, from a Trinitarian faith, "an authentic natural theology is concerned with the discernment of the meaning of life, as much as the demonstration of rationality in faith."[42] Faith is neither credulity nor without a firm foundation for this life and beyond.

In his 2009 Gifford Lectures, *A Fine-Tuned Universe: The Quest for God in Science and Theology*, McGrath again concerns himself with natural theology. He is convinced that "there is growing sympathy for the view that natural theology can provide a deeper understanding on fundamental issues such as the fine-tuning of the universe."[43] While this fine-tuning proves nothing "it is nonetheless deeply suggestive."[44] The question then is how to make sense of these phenomena. To that effect McGrath goes through the different phenomena and their interpretations past and present from the standpoint of natural theology. His main concern thereby is to show the different facets of fine-tuning in the evolutionary process which gave rise to our world and to us within it. While according to McGrath, William Paley has offered a "static vision of natural theology" since he appealed to rational intelligibility, McGrath "finds a new sense of wonder in the vast, complex processes which brought them about."[45] Natural theology is no longer about proving the tenants of the Christian faith, but "the Christian faith, grounded ultimately in divine self-revelation, illuminates and interprets the natural world." The "Book of Scripture" therefore allows for a deeper reading of the "Book of Nature." While natural theology is no longer about proving anything, it "is about a theologically grounded quest for truth, beauty, and goodness within nature."[46] While McGrath is quick to concede that there is a bewildering complexity in nature, he still asserts that "Christian theology provides us with a conceptual net

---

[42] McGrath, *Darwinism and the Divine*, 289.

[43] Alister McGrath, *A Fine-Tuned Universe: The Quest for God in Science and Theology. The 2009 Gifford Lectures* (Louisville, KY: Westminster John Knox, 2009), ix.

[44] McGrath, *A Fine-Tuned Universe*, xiii.

[45] McGrath, *A Fine-Tuned Universe*, 218, for this quote and the next.

[46] McGrath, *A Fine-Tuned Universe*, 219, for this quote and the next.

to throw over our experience of the world, allowing us to make sense of its unity and live with its seeming contradictions." This effort to makes sense of the world by perceiving it through the lens of theology bears much merit. However, we should not forget the struggles Charles Darwin went through when he perceived the "cruelty and mercilessness" in God's good creation. Barth's separation between creation and nature is certainly not helpful in the dialogue with today's enlightened humanity. But all disclaimers notwithstanding, will a natural theology not always be prone to overextend its possibilities to make sense of the world? Would not a theology of nature be more appropriate to interpret the world around us? While a natural theology always seems to imply that we work our way up to God, a theology of nature explicitly starts with God's self-disclosure in Jesus Christ. From this vantage point we are then able to perceive nature as God's creation in spite of all its ambiguity.

CELIA DEANE-DRUMMOND received a PhD in plant physiology in 1980 and a PhD in theology in 1992. She was Professor of Theology and the Biosciences (2000-2011) at the University of Chester, Great Britain, and then was called to serves as Professor in Theology at the University of Notre Dame, with a concurrent appointment with teaching responsibilities in the College of Science (2011-). Informed by her Roman Catholic background she is engaged in constructive systematic theology focusing on the relationship between creation and the natural world. For instance in her 2008 publication *Eco-Theology* (London: Darton, Longman and Todd) she reviews the different Christian approaches to eco-theology shunning the relativism of postmodernity. Though environmental problems have a substantial and reasonable basis in reality, Deane-Drummond is convinced that the sciences are not sufficient to solve environmental problems. Therefore she wants to uncover the theological basis for a proper relationship between God, humanity, and the cosmos. Humans are created but alienated from the world by their tendencies to domination. Yet the story of the world and of humans is one, since the world is our home provided by God.

In her 2004 publication, *The Ethics of Nature* (Oxford:

Blackwell) she emphasizes that we are apart from nature as well as a part of nature. An ethics of nature therefore does not just include how we treat the nature around us but must include who we are as part of nature, as persons. By considering who we are we can gain clues about relating to the world in an appropriate and responsible way. It is interesting that Deane-Drummond dislikes the term stewardship since for her this implies condescendence and instead prefers with Thomas Aquinas virtue and practical wisdom or prudence. Celia Deane-Drummond writes: "I will argue for a virtue ethics approach to complex questions in ecology, drawing specifically on the classical tradition of Thomas Aquinas and including the virtues of wisdom, prudence, justice, and fortitude alongside the theological virtues of faith, hope, and charity."[47] In so doing she covers the whole range of nature, such as animal ethics, biotechnology, cloning, and also feminist approaches to nature and finally arrives at an ethics of wisdom. – Providing the ground for a public theology with regard to climate justice Deane-Drummond again pursues a virtue approach, since "virtues are not just individualistic, that is, they are expressed at different social levels, including the level of the individual, family, civic society and political structural level. ... The virtues are also oriented towards the common good."[48] Whether in a thoroughly secular world the appeal to virtue is sufficient to challenge and move societies toward a sustainable future remains to be seen. Yet Deane-Drummond seems to argue from and within a Judeo-Christian context where such appeal may still get a hearing.

Deane-Drummond explains that according to Thomas prudence or practical wisdom is the "mother" of all other cardinal virtues and it opens up "an approach to decision making at broader, communal levels."[49] Prudence is only possible through

---

[47] Celia Deane-Drummond, "Theology, Ethics, and Values", in *The Oxford Handbook of Religion and Science*, 891.

[48] Celia Deane-Drummond, "Public Theology as Contested Ground: Arguments for Climate Justice", in Celia Deane-Drummond and Heinrich Bedford-Strohm, eds., *Religion and Ecology in the Public Sphere* (London: T & T Clark, 2011), 204-5.

[49] Deane-Drummond, "Theology, Ethics, and Values", 899.

the grace of God and must the seen in the light of the three theological virtues of faith, hope, and charity. Deane Drummond concludes: Virtues "which can be learned through education and family life, become reinforced and transformed into gifts in the context of Christian community, but such community needs to be reminded continually of its dependence on both the grace of God and the wider biotic community in which it is situated."[50]

In *Christ and Evolution: Wonder and Wisdom* (Minneapolis: Fortress, 2009) Deane Drummond intimately interweaves Christology with evolution. Following an overview of ideas and concepts in evolution which are profitable for an engagement with Christology she relates in a constructive way Christology with evolution. She makes generous use of the Swiss Roman Catholic theologian Hans Urs von Balthasar (1905-88) who showed the importance of theo-drama in Christological reflection and also of the Russian Orthodox theologian Sergii Bulgakov (1871-1944) using Sophia (Wisdom) as a foundation for reflecting on incarnation. Moreover, it is a 'deep incarnation' that connects Christ with an evolving creation and indeed the whole cosmos. Deane Drummond is convinced "that Balthasar's discussion of the analogy of being, alongside his consideration of the experience of Christ on Holy Saturday, opening out to affirmation of the Eastern tradition of cosmic Christology, can be appropriated to the specific discussion of the relationship between Christ and nature, and then to Christ and nature understood in evolutionary terms."[51] Since Deane-Drummond emphasizes a cosmic Christology she does not confine herself to Christ's significance for humanity alone but attempts "to think more carefully about the significance of the atonement for the nonhuman world before moving too quickly to considerations about redemption."[52] She does not wish to leave behind the traditional problematic of human sin, but in order to expand the efficacy of Christ's atoning work to the nonhuman world she

---

[50] Deane-Drummond, "Theology, Ethics, and Values", 904.

[51] Celia Deane-Drummond, *Christ and Evolution: Wonder and Wisdom* (Minneapolis: Fortress, 2009), 129.

[52] Deane-Drummond, *Christ and Evolution*, 159.

must first connect human sinfulness to the rest of creation. She does this in two ways. First, she connects human sin to our evolutionary past by suggesting that there are nonhuman forms of animal morality. Influenced by the work of Dutch-American primatologist Frans de Waal (*1948) and others, Deane-Drummond concludes that some nonhuman animal species can be said to have their own form of morality measured by their capacity for flourishing in their own worlds. Second, she connects human sin to our increasing potential for "anthropogenic evils (evils suffered in the nonhuman sphere as a result of human activity)."[53]

In 2012 she delivered the Boyle lecture on "Christ and Evolution: A Drama of Wisdom?" at St. Mary LeBow in London in which she rephrased much of this book. There she concludes form Christ's atoning work of self-sacrifice and obedience to God:

> If other human beings choose to follow this pattern, then they would try and perceive goodness through the crystal lens of truth set forth by the purity of Christ's manner of living and dying and rising again. … When we reflect on the tremendous practical ecological and social problems facing our own generation, many of these have tragically been of our own human making. However, the hope that the Christian faith in Christ can inspire is one that affirms self-destruction and that of our world need not be the final act in the theo-drama that interweaves both human and creaturely life.[54]

Then she ends with an eschatological note opting for the coming of Christ as our source of hope. It is clear that she does not advocate some kind of transhumanism to pave a new future for humanity. Rather she concludes: "Human responsibility is less about seeking new evolutionary pathways for itself than it is a matter of engaging in practices that will help sustain the biodiversity on which human life as we know it depends. Transhumanism represents a callous escape from that responsibility

---

[53] Deane-Drummond, *Christ and Evolution,* 174.

[54] Celia Deane-Drummond, "Christ and Evolution: A Drama of Wisdom?", in *Zygon* (September 2012) 47:538, in her revised and expanded version of her Boyle Lecture.

and a severing from evolutionary history in the guise of the rhetoric of evolutionary psychology."[55] Honoring her religious tradition Celia Deane-Drummond presents a sometimes highly intricate interweaving of scientific and theological insights.

A person to watch for is ANNA IJJAS. Her 2010 PhD dissertation in Roman Catholic theology at the University of Munich was published under the title *Der Alte mit dem Würfel* (*The Old One with the Dice*). She is a member of the *International Society for Science and Religion* and was a doctoral student at the Max Planck Institute for Gavitational Physics (Albert Einstein Institute) in Potsdam, Germany, working on string cosmology. She has also been a Thyssen research fellow at the Center for Astrophysics, Harvard University. Presently she is a postdoctoral fellow at the Princeton Center for Theoretical Science doing research in theoretical cosmology. In her dissertation she starts with the exchange between Max Born (1882-1970) and Albert Einstein, in which the latter claimed that God does not play dice. Ijjas wants to refute Einstein's assertion with the help of quantum mechanics and also show "that a dialogue between physics and metaphysics, between theology and science is both possible and necessary."[56] Yet she knows that "it seems as if quantum mechanics were to some extent a theory for everything."[57] It is hard to think of any metaphysical problem that has not been – allegedly – solved by applying the principles of quantum theory.

In her book, Ijjas investigates whether and to what extent quantum theory can usefully contribute to the resolution of fundamental theological and philosophical dilemmas. To enable her to do this, she has developed a new methodological approach which allows her critically to probe the links between quantum mechanics and metaphysics. She is particularly interested in the question of whether the theoretical foundations of quantum mechanics can be logically reconciled with various metaphysical

---

[55] Deane-Drummond, *Christ and Evolution,* 286 f..

[56] Anna Ijjas, *Der Alte würfelt nicht. Ein Beitrag zur Metaphysik der Quantenmechanik* (Göttingen. Vandenhoeck & Ruprecht, 2011), 9.

[57] Anna Ijjas, *Der Alte würfelt nicht,* 15.

models – such as the concept of determinism. She realizes that one can show that "the statistical character of quantum mechanics could be used as an argument for indeterminism."[58] With regard to the issue of a free will indeterminism in quantum mechanics appears to be very significant. "The discussion which was declared closed by leading brain researches can now be opened again anew."[59] Ijjas asserts that "with quantum physics a theist gains an empirical argument for the claim of free will and for the reconcilability of theism with the Darwinian theory of evolution."[60] She concludes: "I believe that Einstein was wrong."[61] The structure of the universe allows for a degree of open-endedness in physical processes. But quantum physics gives us no grounds for believing that the world is dominated by blind chance. In this way Einstein was correct. There is freedom in creation and therefore also the risk of evil and sorrow which we experience every day.[62]

A very different dialogue partner is the Dutch theologian and philosopher of religion WILLEM DREES whom we have already met as the editor of *Zygon*. Besides many other prestigious tasks he served as president of *ESSSAT, The European Society for the Study of Science and Theology*. Similar to Polkinghorne he studied natural science, more specifically theoretical physics, and also theology. Willem Drees confesses that he pursues a "monistic train of thought."[63] For him there is no knowledge which is not of scientific nature and inaccessible to scientific investigation. He

---

[58] Anna Ijjas, *Der Alte würfelt nicht*, 148.

[59] Anna Ijjas, *Der Alte würfelt nicht*, 186.

[60] Anna Ijjas, *Der Alte würfelt nicht*, 205.

[61] Anna Ijjas, *Der Alte würfelt nicht*, 207.

[62] In a more recent essay Ijjas convincingly shows that quantum theory can contribute to reconcile evolutionary biology with the claim that the universe was created by a loving creator. See Anna Ijjas, "Quantum Aspects of Life: Relating Evolutionary Biology with Theology via Modern Physics," in *Zygon* (March 2013), 48:60-76.

[63] Willem B. Drees, *Vom Nichts zum Jetzt: Eine etwas andere Schöpfungsgeschichte*, trans. from Dutch Klaus Blömer (Hannover: Lutherisches Verlagshaus, 1998), 91.

rejects the claim that theology is "the science of God."[64] Rather he admits "that we have no knowledge of God."[65] With this admission God and faith are not done away with but Drees claims that it is necessary to speak dualistically. "In each moment there is more than that which one knows about humanity. There is more than facts; values and visions are a reality as values and visions of humans." Therefore the statement that in Jesus God's intentions, that is God's love and forgiveness, have been disclosed, is a judgment based on certain facts. Drees regards "theology as an interpretation of human existence with the help of a certain religious tradition" in which "the factual and the normative elements are interconnected."[66] Since the scientifically graspable reality is not self-evident and our grasping of this reality cannot explain why this reality exists Drees refers to the metaphor of God as the creator, however not in an anthropomorphic way but as "the ground of all being."[67]

The naturalistic worldview of Drees reminds us of Albrecht Ritschl and his distinction between factual judgments and value judgments. The natural sciences deal with facts, religion and faith deal with values. Drees emphasizes this distinction again firmly in *Beyond the Big Bang* a publication in which he deals extensively with the dialogue between theology and the natural sciences. He writes in the preface: "I therefore develop an outline of a theology which takes science seriously, but does not restrict itself to the quest for a fit with the results of science. I hold that an adequate theology should deal with experiences of imperfection and injustice, and hence has to maintain a 'prophetic' dimension, a judgment of disparity between the way things are and the way they should be."[68]

---

[64] Willem B. Drees, *Creation from Nothing until Now* (London: Routledge, 2002), 54. This publication is a considerably rev. ed. of *Vom Nichts zum Jetzt*.

[65] Willem B. Drees, *Vom Nichts zum Jetzt*, 92 f., for this and the following quotation.

[66] Drees, *Vom Nichts zum Jetzt*, 70.

[67] Drees, *Vom Nichts zum Jetzt*, 106.

[68] Willem Drees, *Beyond the Big Bang: Quantum Cosmologies and God* (La Salle, IL: Open Court, 1993 [1990]), xiii.

In this work, originally his dissertation in theology (Groningen 1989), Drees also pursues "a scientific agnosticism with respect to that transcendence" because one can consider the universe on the one hand as a free gift of an ontological transcendent creator or on the other hand as something that simply is.[69]

Since our knowledge is limited in its reach and in its conceptuality one cannot know God either through ordinary experience or through the natural sciences. According to Drees this is also not necessary because "moral and ascetic values inhabit a domain of reality other than scientific knowledge."[70] But Drees is concerned about such values because if one claims that God is the source of reality then ultimately the world is affirmed as good. It is not recognized as such but affirmed. Through this there is an active role of humanity: "As we affirm the integrity of creation as a value and as a theologically postulated reality, we commit ourselves to contribute to that integrity."[71] Theology is concerned with postulates of value and natural science with the recognition of reality. Therefore one can even talk about an agreement, a consonance between theology and the natural sciences. This, however, is a human construct and is not found in reality itself. That is not only the case because religious convictions cannot be proven on the basis of scientific theories but also because science is always in flux. A theologian must find a dialogue partner who represents the present scientific consensus but who comes from a different "culture" than theology.[72]

In a more recent publication, *Religion, Science, and Naturalism* Drees affirms: "This book takes a more radical naturalist position than most on religion and science."[73] Nevertheless he now admits: "Religion is about reality, that is about creation and

---

[69] Drees, *Beyond the Big Bang*, 11.

[70] Drees, *Beyond the Big Bang*, 110.

[71] Drees, *Beyond the Big Bang*, 209.

[72] Willem Drees, *Religion, Science and Naturalism* (Cambridge: Cambridge University, 1996), 239, where he also rejects the idea to understand scientific knowledge "ahistorically."

[73] So Drees, *Religion, Science and Naturalism*, prior to the title page.

about God. Science informs us about reality."[74] Therefore it is even conceded to religion that it deals with reality. Since in the world everything happens in a totally natural way Drees does not want to understand God's activity in the unpredictability of complex processes or the way the behavior of particular constituents in a system is shaped by the state of the system as a whole, but he distinguishes "between divine action (as atemporal creation of the whole) and the temporal processes in the world."[75] As Aristotle had already claimed, God therefore is the first unmoved mover who set everything in motion or, by way of classical deism, the creator God then left the world to run by itself.

Though in his naturalistic view for Drees the natural sciences play the main part, they are not the only undertaking which is concerned with knowledge. This is also a religious view of reality. Religions with their rituals and myths have arisen in particular environments and were formed by the challenges which humans had to meet. But religion is not simply the result of evolution. A religious tradition "via its metaphors, concepts, and images evokes a conception of moral and spiritual good life."[76] This does not mean according to Drees that religion is only a practical undertaking. It orients itself according to ultimate ideals which transcend any actually attainable goal or any attainable situation. Since religion and theology reflect religion which concerns itself with reality, but in a very different way than is done by the natural sciences, there can be no conflict if theology and the natural sciences are mindful of their different way of approaching the one reality. There is only meaningful supplementation. From his naturalistic presupposition it is understandable why Willem Drees became the new editor of *Zygon* a journal that dates back to Ralph Wendell Burhoe, a Unitarian. Drees himself affirms that he comes out of liberal Protestantism to which he owes much. With this he distinguishes himself from Burhoe but also from theologians in the Germanic region who concern themselves with the dialogue between theology and the natural sciences.

---

[74] Drees, *Religion, Science and Naturalism*, 92.

[75] Drees, *Religion, Science and Naturalism*, 94.

[76] Drees, *Religion, Science and Naturalism*, 276.

8.2 Dialogue as an Avocation (Moltmann, Pannenberg)

As mentioned previously, there are mostly the classical disciplines in the German university system represented by professorships, for example, in dogmatics and fundamental theology on the Roman Catholic side and systematic theology on the Protestant side. But there are no special chairs or professorships for theology and the natural sciences as is the case with WENTZEL VAN HUYSSTEEN (*1942) at Princeton Theological Seminary, or whole institutes such as the *CTNS* in Berkeley. Even the institute *Technology, Theology, and the Natural Sciences* (*Technik, Theologie, Naturwissenschaften; TTN*) at the Ludwig Maximilian University in Munich is not headed by a professor but by a scholar appointed and paid by the Lutheran Church. Therefore it is not surprising that most German theologians who show a concern for the dialogue between theology and the natural sciences do so only with support of occasional designated contributions. But this does not mean that they treat the dialogue with the natural sciences with less seriousness than their colleagues who can spend most of their professional life in this arena.

For instance the Heidelberg systematician MICHAEL WELKER (*1947) published in 1995 *Schöpfung und Wirklichkeit* (Neukirchener Verlag; Eng. trans. *Creation and Reality*, Minneapolis: Fortress, 1999) in which he argued for a biblical theology and against a natural theology. In the *Handbuch Systematischer Theologie* (*Compendium of Systematic Theology*) CHRISTIAN LINK (*1938) published a two-volume theology of creation in which he subtitled volume 2: *Schöpfungstheologie angesichts der Herausforderungen des 20. Jahrhunderts* (*Theology of Creation Facing the Challenges of the 20ᵗʰ Century*; Gütersloh: Gerd Mohn, 1991). There he dealt extensively with the interaction between theology and the natural sciences and considered time as the common horizon of both. "Theology and the natural sciences move in the common horizon of time."[77] Since the natural sciences are only

---

[77] Christian Link, *Schöpfung: Schöpfungstheologie angesichts der Herausforderungen des 20. Jahrhunderts*, vol. 7/2: *Handbuch Systematischer*

concerned about one mode of time, the present, they can only discover one partial realm of the phenomena. Theology however through the resurrection goes beyond all available boundaries which have arisen from the past and the present and arrives at a different point of view. Yet for Link there is still "a complementarity of two incompatible points of view." It is difficult to understand how one can still talk with Link about a complementarity.

ULRICH EIBACH (*1942) has entered the dialogue with the natural sciences much more extensively, but here mostly from an ethical angle for instance with topics such as euthanasia, embryonic research, and neurobiology. For him the dialogue is no theory because he writes with the experience of a clinical counselor and from his conversations with nurses and medical doctors in hospitals and with biologists in their research. His perspective is from a conservative Christian view of humanity.[78] – One should also not omit the Heidelberg systematician JÜRGEN HÜBNER who still works at the *FEST* and who has contributed much to a historical understanding of the natural sciences from a theological perspective.[79]

Parenthetically I also want to mention that I occasionally concern myself with the dialogue between theology and the natural sciences, once through the editing of sixteen volumes of the yearbook of the Karl-Heim Society *Faith and Thought* (*Glaube und Denken*), and in various articles, but especially in the publication of my book entitled *Creation* (Grand Rapids: William B. Eerdmans, 2002)[80]. It is especially important for me to

---

Theologie (Gütersloh: Gerd Mohn, 1991), 454, for this and the following quotation.

[78] Cf. Ulrich Eibach, *Sterbehilfe – Töten aus Mitleid* (Wuppertal: R. Brockhaus, 1998; *Gentechnik und Embryonenforschung – Leben als Schöpfung aus Menschenhand?* (Wuppertal: R. Brockhaus, 2005 [2002]); *Gott im Gehirn – Ich eine Illusion? Neurobiologie, religiöses Erleben und Menschenbild aus christlicher Sicht* (Wuppertal: R. Brockhaus, 2006, 3rd ed. 2010).

[79] Jürgen Hübner, *Die Theologie Johannes Keplers zwischen Orthodoxie und Naturwissenschaft* (Tübingen: Mohr/Siebeck, 1975); Jürgen Hübner et. al., eds., *Theologie und Kosmologie: Geschichte und Erwartungen für das gegenwärtige Gespräch* (Tübingen: Mohr/Siebeck, 2004).

[80] The German version was published with the title *Schöpfungsglaube im*

show in this dialogue how in a world shaped by the natural sciences and technology one can believe in God the creator, sustainer, and redeemer of the world and of humanity without sacrificing one's own intellect. Already earlier I elaborated on this concern in an anthropology which I have recently updated and considerably expanded.[81] Yet I have never seen this dialogue, necessary and important as it is, as my main task in theology. This is also true for the Tübingen systematician Jürgen Moltmann.

With JÜRGEN MOLTMANN (*1926) there prevails an ecologically colored understanding of nature in creation because for him it is important to develop an ethics for the scientific technological mastery of the world. Yet how this cooperation between theology and the natural sciences should take shape is not extensively treated in his earlier publications. Instead Moltmann points out that modern industrial society will exhaust the resources of nature and the ruthless exploitation of the natural resources will destroy the basis for life.[82] Yet Moltmann does not consider an ecological catastrophe as unavoidable since it has been encouraged by a dubious interpretation of creation. "For centuries, men and women have tried to understand God's creation *as nature,* so that they can exploit it in accordance with the laws science has discovered. Today the essential point is to understand this knowable, controllable and usable nature *as God's creation,* and to learn to respect it as such."[83] Moltmann calls here for a different way of thinking whereby on the one hand nature is freed from its

---

Horizont moderner Naturwissenschaft (Neukirchen-Vluyn: Friedrich Bahn-Verlag, 1996).

[81] Hans Schwarz, *Our Cosmic Journey: Christian Anthropology in the Light of Current Trends in the Sciences, Philosophy, and Theology* (Minneapolis: Augsburg, 1977), and Hans Schwarz, *The Human Being: A Theological Anthropology* (Grand Rapids, MI: Wm. B. Eerdmans, 2013).

[82] Vgl. Jürgen Moltmann, *Gerechtigkeit schafft Zukunft: Friedenspolitik und Schöpfungsethik in einer bedrohten Welt* (Munich: Christian Kaiser, 1989), 70.

[83] Jürgen Moltmann, *God in Creation: A New Theology of Creation and the Spirit of God. The Gifford Lectures 1984–1985,* trans. Margaret Kohl (Minneapolis: Fortress, 1993), 21.

domination by humanity and on the other hand humanity is freed from its anti-natural striving for power and thereby reaches its natural communion with the other creatures of God. In this way Moltmann develops a doctrine of creation from an eschatological or rather messianic perspective.

It was not until 2002 in a publication which essentially consists of articles published previously that he explicitly focuses on the dialogue and emphasizes: "From very early on, the theological discussion with scientists fascinated me."[84] Yet he knows how difficult it is to bring into dialogue pure natural science and scholarly theology because most representatives do not see a platform on which they can communicate with each other. But when they get together in "commissions for ethics in the sciences" then potential or factual results of research are already presupposed and the ethical reflection comes too late. "Scientific research is objective, but it is not value-free. It is subjected to the utilitarian interests of society."[85] One obtains money for research only if there is a societal, political, or industrial interest in that research.

Moltmann suggests that theology should not only take the book of the Bible seriously but also the book of nature. Through natural theology which is derived from the book of nature "people become *wise* in their dealings with nature; but they would not be saved. Salvation is given only through perception of the revelation of God the redeemer."[86] Through the search for wisdom one discovers and learns. Thereby theologians can enter into conversation with natural scientists. The guiding question in recognition would no longer be what one can make from an object which one has explored or how one can transform it usefully for humans. This human claim for domination only brings nature to silence and it makes humans dumb instead of wise. Rather, a dialogue should ensue between discovered natural wisdom and human wisdom which is to be learned. "Human wisdom will search for viable harmonizations between human civilization and the earth's eco-

---

[84] Jürgen Moltmann, *Science and Wisdom*, trans. Margaret Kohl (London: SCM 2003), x (in his preface).

[85] Moltmann, *Science and Wisdom*, 38 f.

[86] Moltmann, *Science and Wisdom*, 27.

systems. Then the goal is not human domination over nature; it is a well-adjudged and prudent conformity with nature."[87] It is surprising how Moltmann has overcome the Barthian narrowness and again advocates taking seriously the book of nature. This approach from the side of theology would be a good beginning for the dialogue. But also for the natural sciences is ultimately the important basic question "How do I get money for my research?" remembering that research is connected with wisdom and not only with objective knowledge.

WOLFHART PANNENBERG (*1928) has also not made the dialogue between theology and the natural sciences his main topic. Yet from the very beginning it was important for him to represent the Christian faith in a credible manner because, in following his mentor Hegel, reason and revelation for him are not opposites. Already in his "Dogmatic Theses on the Doctrine of Revelation" (1961) he rejected a special salvation history and placed theology in the same realm as any other science. He claimed that historical revelation is open to everyone and therefore one can understand the biblical reality of God's self-disclosure.[88] In this way Pannenberg opens up his theological work to the impact of science and is ready to allow for falsification on the basis of science.[89] Pannenberg wants theology to be included — as an equal partner — in the community of rational scientific investigation of the world. In anthropology, for instance, his goal is "to lay theological claim to the human phenomena described in the anthropological disciplines. To this end, the secular description is accepted as simply a provisional version of the objective reality, a version that needs to be expanded and deepened by showing that the anthropological datum itself contains a further and theologically

---

[87] Moltmann, *Science and Wisdom*, 29.

[88] Cf. Wolfhart Pannenberg, "Dogmatic Theses on the Doctrine of Revelation" (thesis 3), in *Revelation as History*, ed. Wolfhart Pannenberg, trans. D. Granskau (New York: Macmillan, 1968), 135.

[89] So Philip Hefner, "The Role of Science in Pannenberg's Theological Thinking," in Carl E. Braaten/Philip Clayton, eds., *The Theology of Wolfhart Pannenberg* (Minneapolis: Augsburg, 1988), 284.

relevant dimension."[90] Pannenberg does not simply isolate certain facts and make them fruitful as a foundation for revelation. He goes further. He is convinced that "when the events of nature and history events are properly understood, in and of themselves, knowledge of their being rooted in God and God's will is conveyed."[91] But complete knowledge is only available once history has come to its end.

In his comprehensive treatise on the theory of knowledge, Pannenberg endeavors to show the scientific character of theology. He claims that theology is a science of God which can approach its object matter only indirectly through the study of religions. To fulfill its scientific character, theological assertions must meet three criteria:

First, they must "have a cognitive character"; this means that they must "say something about a state of affairs for which they claim truth."[92]

Second, these assertions must be coherent. They have to refer to one object matter. This object matter is given in the indirect self-communication of the divine reality. This divine self-communication occurs in the preliminary experiences of the totality of the reality of meaning as they are present in the religious traditions of faith.

Finally the assertions must be open to examination. Pannenberg introduces a preliminary verification about the truth claim. A final verification cannot be established either positively or negatively within the process of history as long as this process is not yet closed. This means that even the assertions of truth in the natural sciences are preliminary and not essentially different from theological claims of truth.

In his 1970 essay "Contingency and Natural Law," Pannenberg emphasizes that a common ground upon which natural science

---

[90] So Wolfhart Pannenberg, *Anthropology in Theological Perspective,* trans. Matthew J. O'Connell (Philadelphia: Westminster, 1985), 19 f.

[91] Philip Hefner, "The Role of Science in Pannenberg's Theological Thinking," 269.

[92] Wolfhart Pannenberg, *Theology and Philosophy of Science,* trans. Francis McDonagh (London: Darton, Longman & Todd, 1976), 327.

and theology can meet without denying their specific differences is that of contingency and order.[93] The Judeo-Christian understanding of God was always decisively influenced by contingent historical events. "New and unforeseen events take place constantly that are experienced as the work of almighty God." One also recognized an orderliness which was dependent on the contingent activity of God.

Another area in which Pannenberg touches the natural sciences is with the concept of "field." Especially in developing his understanding of the doctrine of God as spirit does he use the concept of field as it is commonly understood in physics and applies it to theology.[94] Pannenberg regrets that spirit is traditionally associated with intellect. In the Old Testament, however, the term "spirit" more frequently denotes a breeze or wind. By equating the spirit of God with the human spirit, there occurred an excessive anthropomorphism in understanding the divine reality. To counteract this tendency Pannenberg takes over the concept of field. According to Pannenberg,

> spirit is rather a kind of force, comparable to the wind, but prior to bodily phenomena. If theology wants to be true to the biblical witness, the concept of God as spirit has to be disentangled from the customary identification with mind, an identification which entails an all-too-facile image of God as 'personal.'

When Pannenberg specifically refers to creation, he again uses the concepts of contingency and field. The theological assertion that the world is contingent on an act of divine creation implies, according to Pannenberg, the assertion "that the existence of the

---

[93] Cf. Wolfhart Pannenberg, "Contingency and Natural Law," in Pannenberg, *Toward a Theology of Nature: Essays on Science and Faith,* ed. Ted Peters (Louisville: Westminster/John Knox, 1993), 76, including the following quote.

[94] Cf. for the following Wolfhart Pannenberg, "Theological Appropriation of Scientific Understandings: Response to Hefner, Wicken, Eaves, and Tipler," *Zygon* 24 (1989): 256 f.; quote on 258.

world as a whole and of all its parts is contingent."[95] The world as a whole need not exist at all, but it owes its existence to the free activity of divine creation. This is also true for each part of the world. The contingency of the world shows a close connection between creation and its preservation. The world was not just once called into existence, but each created creature must be preserved in every moment of its existence if it is not to perish. In the Christian tradition such preservation is nothing but continuous creation. Creation occurred not just at the beginning but is verified anew in each moment.

Pannenberg uses field theory, for instance, to translate assertions about angels into our present-day conceptual world.[96] Traditionally angels are considered to be immaterial spiritual realities and powers which, in distinction to the divine spirit, are limited realities. They either verify themselves as God's messengers or they oppose God in demonic freedom. In a field structure angels could be interpreted as the appearance of relatively independent parts of a cosmic field. If one regards the background of the biblical language which talks about angels as personal realities, then one could very well think of them as fields of power or dynamic spheres. As such they could be experienced either as good or evil. Of course, Pannenberg knows that with these deliberations he deviates from the traditional doctrine of angels, but with these metaphors he attempts to grasp that which is theologically essential when we speak about angels. Especially through his use of the field theory Pannenberg received strong criticism because the field theory is understood completely different from the way Pannenberg uses it. He does not use it as a metaphor but as a (theological) concept.[97]

Pannenberg is convinced that "the theological doctrine of creation should take the biblical narrative as a model in that it

---

[95] Wolfhart Pannenberg, "The Doctrine of Creation and Modern Science," *Zygon* 23 (1988): 8.

[96] For the following, Pannenberg, "Doctrine of Creation," 14 f.

[97] Cf. Mark Worthing, *God, Creation, and Contemporary Physics* (Minneapolis: Fortress, 1996), 120–25, who criticizes that Pannenberg depends too much on a concept in physics.

uses the best available knowledge of nature in its own time in order to describe the creative activity of God. This model would not be followed if theology simply adhered to a standard of information about the world which has become obsolete long ago by further progress of experience and methodical knowledge."[98] Pannenberg wants to express faith in creation, with the help of modern scientific knowledge, in such a way that the biblical narrative only provides a model for that which ought to be expressed. The content of our faith is thereby largely filled in by our contemporary knowledge. In so doing he can be critical, for instance, with respect to the Priestly creation account which places the creation of the stars relatively late. At the same time he notes astounding analogies in the first chapter of the Bible between our present ideas of the origin and development of the world with that of the people of antiquity. Important for him is not so much the *how* of creation, something that, considering the relativity of its knowledge, modern science can express best, but the *that* of creation. Therefore he does not discuss the assertions of an alternative creationistic science, but simply states: "Theology has to relate to the science there *is* rather than invent a different form of science for its own use."[99] Pannenberg totally relies on the results of modern science. He is confident that truth cannot be divided and that there is no opposition between theology and the natural sciences.

---

[98] Pannenberg, "Doctrine of Creation," 19.
[99] Pannenberg, "Doctrine of Creation," 7.

9. Important Issues

After our extensive review of the most important dialogue partners in the discourse between theology and the natural sciences we briefly want to touch on a few issues which are especially important and necessary for the dialogue. There are on the one hand the classical issues which are concerned more with the theory of knowledge stemming from cosmology and biology and also the question concerning creation and its creator. On the other hand there are new and practical issues which the applied sciences pose to us, such as in brain research and in medicine. After these great advances in knowledge one asks here about the function of the brain and whether there is still a free will and given the immense progress in medicine whether to use the pre-implantation diagnostics just to mention a few controversial issues in the present-day discussion.

9.1 Nature or Creation?

The alternative of nature or creation implies at the same time the question about God. Is our world the result of purely natural causes or is there a higher power active behind it, for instance of a benevolent God? If the latter were true we should at least be able to discern something of God's activity in nature. With this assertion we are immediately confronted with the often emotionally discussed issue of a design in nature.

9.1.1 Creator and Creation

Natural scientists and lay persons in science today confront the same problem which also vexed Darwin: in his observations he concluded that life develops in a natural way through accidental variations and subsequent selection. Nevertheless he could not believe that this was everything. This ambivalence of Darwin is resolved by Dawkins and others when they conclude that everything which exists is a physical unity which can ultimately be fully explained with the laws of physics. Even humans are only a very complicated biological "machine." This option is usually called *physicalism*. Others, such as John Eccles, advocate a more dualistic viewpoint. It says that in humans and perhaps also in other closely related species that next to physical components there are also non-physical ones which can be designated as soul, spirit, or the self and are not reducible to physical components. This second option is usually termed *dualism*. In more recent time a third option was posited, the so-called theory of *emergence*.[1] From the theological perspective it could be concluded that God functions with a finite living being as co-creator without exerting any force and without preempting the result which this co-creation causes. For this process God established the external parameters which are now discovered by the natural scientist. God makes it possible through the open process of evolution that ever more complex organisms and systems that react with each other are constructed so that eventually those living beings emerge as we are. God acts invisibly and unnoticeably behind and through the natural processes.

The question of how such a God is distinguished from no God at all is answered by the theologian and philosopher of religion PHILIP CLAYTON (*1956). He claims "that the world has not turned out to be explainable solely in terms of the matter and energy relations of physics as we know it."[2] He asks us to consider

---

[1] Cf. to this point Philip Clayton, *Mind and Emergence. From Quantum to Consciousness* (Oxford: University Press, 2004), who introduces in his preface these three options (p. v) and claims that the theory of emergence is to be preferred to the two other options.

[2] So Philip Clayton, "Emergence from Quantum Physics to Religion: A

that the universe could be regarded as a system closed in and of itself. But most smaller systems are thermodynamically open and therefore perishable and contingent. Furthermore one must consider the uncertainty on the quantum level which in some cases has macro-physical effects. There is a spontaneity here or at least an unpredictability which has results on our world of experience. With chaotic systems one cannot exactly predict the future because to do so one would have to measure precisely its initial state, which is empirically impossible. This means there are limits to our scientific knowledge. God who is made possible through these considerations resembles very much on the one hand the God of Hawking who determined the initial parameters after which everything ran its own course. But one could here also think of a God who acts outside the limits of our scientific possibility of knowledge. If scientific knowledge changes then this view of God would also have to be revised.

Arthur Peacocke attempted to dispel these anxieties by starting with an all-knowing God who knows both the world in its totality and also all its parts, structures, and processes.[3] For Peacocke there is this panentheistic perspective in which God, so to speak, comprises the whole world. Since God is "larger" than the world, just like a world logos or a world reason, God can call into existence parts of creation. "God could cause particular events and patterns of events to occur which express God's intentions. These would then be the result of 'special divine action' as distinct from the divine holding in existence of all-that is, and would not otherwise and have happened had God not so intended." Therefore the conformity with the laws of nature is not abolished but within these there is an openness which shows itself

Critical Appraisal", in Philip Clayton/Paul Davies, eds., *The Re-emergence of Emergence: The Emergenist Hypothesis from Science to Religion* (Oxford: Oxford University, 2006), 308, and for the following.

[3] Cf. Arthur Peacocke, "Emergent Realities with Causal Efficacy: Some Philosophical and Theological Applications," in Nancey Murphy/William Stoeger, eds., *Evolution and Emergence: Systems, Organisms, Persons* (Oxford: Oxford University, 2007), 278 f., including the following quotation.

as God's activity. With this view one could easily think of an intervention of God "from above" just as many people today think of God's activity in this world.

For John Polkinghorne this view is too restrictive and he asks us to consider the following: since God "is not embodied in the universe ... there does not seem to be any reason why God's interaction with creation should not be purely in the form of active information. This would correspond with the divine nature being pure spirit and would give a unique character to divine agency in a way that theologians have often asserted to be necessary. (God is not just an invisible cause among other causes)."[4] This way we have found our way back to an almost classical relationship of creator and creation: that the creator is present in creation at every moment but also stands over against it. The world would be freed from a physical determinism because God could shape the world in freedom or in lawfulness.

As can be detected from this discussion which must be continued in the face of the scientific progress in knowledge, there are different possibilities to show that the chains of events which the natural sciences discover in the biological realm do not exclude a divine action in the whole and in detail. This seems also to be the case in the cosmological realm.

9.1.2 Fine Tuning and Design

We have already mentioned the so-called fine tuning which led Tipler and Barrow to their thesis of an anthropic principle. Amazing in this regard is the "conversion" of the British philosopher ANTONY FLEW (1923 – 2010) who once vehemently denied with logical arguments God's existence and was a decided atheist. Yet now he has radically changed his conviction. The picture of the world painted by the modern natural sciences Flew now uses to point to God in a threefold way: "The first is the fact

---

[4] John Polkinghorne, "The Metaphysics of Divine Action," in John Robert Russell *et al.*, eds., *Chaos and Complexity: Scientific Perspectives on Divine Action* (Notre Dame, IN: University of Notre Dame, 1995), 155 f.

that nature obeys laws. The second is the dimension of life, of intelligently organized and purpose-driven beings, which arose from matter. The third is the very existence of nature."[5] Flew claims that confronted with the fine tuning of the universe the atheists now have trouble answering the question of how the universe came into existence. Also the thesis that there are many universes and that only ours has the right laws enabling us to exist is unconvincing for Flew, because the existence of other universes is an assertion which cannot be proven. Yet Flew acknowledges that there is no scientific proof for God either.

> Science qua science cannot furnish an argument for God's existence. But … the laws of nature, life with its teleological organization, and the existence of the universe—can only be explained in the light of an Intelligence that explains both its own existence and that of the world. Such a discovery of the Divine does not come through experiments and equations, but through an understanding of the structures they unveil and map.[6]

Flew distinguishes himself from the *Intelligent Design Movement* which claims to prove scientifically that there must be an intelligence behind the world. He points out that such a creative intelligence is the most plausible explanation of our world and its structure. How controversial his position is, however, becomes evident when his exposition is not considered by some as that of a scholar in his mature years but that of a senile old man. Yet when we review the authors who he approvingly quotes on the inside cover page of his book who agree to him, we realize that there will be further discussion as the knowledge is advanced in both cosmology and biology.

---

[5] Antony Flew with Roy Abraham Varghese, *There Is A God: How the World's Most Notorious Atheist Changed His Mind* (New York: HarperOne, 2007), 88 f.

[6] Antony Flew with Roy Abraham Varghese, *There Is A God,* 155.

9.1.3 Appendix: Design in the Roman Catholic Tradition[7]

While the *Intelligent Design Movement* is problematic for many
people, the design argument is the basis for Flew's thesis and it
has solid anchorage in the Roman Catholic tradition. For instance Cardinal CHRISTOPH SCHÖNBORN (*1945) of Vienna,
Austria, asserted in a guest commentary in the *New York Times*
"that by the light of reason the human intellect can readily and
clearly discern purpose and design in the natural world, including in the world of living things."[8] This is no special idea of
Cardinal Schönborn but he is quoting a statement of Vatican
Council I (1869/70) almost verbatim. There in the *Dogmatic
Constitution on the Son of God* it was stated concerning revelation: "The same Holy Mother Church holds and teaches that
God, the source and end of all things, can be known with certainty
from the things that were created, through the natural light of
human reason."[9] The *Dogmatic Constitution* refers here to Romans 1:20. This statement is, of course, in line with THOMAS
AQUINAS (1225 – 1274) who at Vatican I was pronounced the official teacher of the church. In his *Summa Theologica* Thomas
emphasized that human reason is able to prove the existence of
God. This line of argument was maintained at Vatican II (1963 –
65) where it states in the *Dogmatic Constitution on Divine Revelation:* "This sacred Synod affirms, 'God, the beginning and end
of all things, can be known with certainty from created reality by
the light of human reason' (cf. Romans 1:20); but the Synod

---

[7] Concerning the Intelligent DesignTradition in general see Hans
Schwarz, "Die Intelligent Design Tradition", *theologische beiträge* 40 (June
2009), 152 – 166.

[8] *New York Times* (July 7, 2005). *www.nytimes.com/2005/07/.../
07schonborn.htm* and Robert John Russell, "A Critical Response to Cardinal Schönborn's Concern over Evolution," *Theology and Science* (July
2006), 4:193.

[9] Josef Neuner/Jacques Dupuis, *The Christian Faith in the Doctrinal
Documents of the Catholic Church,* 7th ed. (New York: Alba House, 2001), 43,
and Denzinger/Schönmetzer, *Enchiridion Symbolorum Definitionum et
Declarationum de Rebus Fidei et Morum,* 34th ed. (Rome: Herder, 1968),
588 (3004).

teaches that it is through His revelation 'that those religious truths which are by their nature accessible to human reason can be known by all men with ease, with solid certitude, and with no trace of error'."[10] Naturally, this is referring to the above quoted statement of Vatican I. We detect here an optimistic evaluation of the capabilities of human reason. Human sinfulness has not damaged reason with regard to its capability of knowing God. Not only possibly, but with ease and certainty, God can be known through creation.

With regard to human reason, Roman Catholic theology has always been more optimistic than, for instance, Lutheran theology. Luther would agree with Thomas Aquinas that human reason can "recognize" the existence of God by observing nature. Luther too makes frequent reference to Romans 1:19 ff. and states that nature teaches all people "that there is a God who gives us all things good and helps us against all evil."[11] But such knowledge is always ambivalent and gives us no certainty. Real knowledge about God can only come through God's self-disclosure in Jesus Christ. But is this attitude so different from what the Roman Catholic Church teaches? We read in the *Catechism of the Catholic Church:* "Starting from movement, becoming, contingency, and the world's order and beauty, one can come to a knowledge of God as the origin and end of the universe."[12] Explicit reference is made here to the so-called proofs of the existence of God while at the same time stating that they are "not in the sense of proofs in the natural sciences, but rather in the sense of 'converging and convincing arguments.'" Again, there is reference made to Romans 1:19 f. This means, that nature points to God's existence, but this existence cannot be proven in a strict sense of the term.

In the *Catechism* it then states: "The world, and man, attest that they contain within themselves neither their first principle nor their final end, but rather that they participate in Being itself,

---

[10] *Dogmatic Constitution on Divine Revelation* (6).

[11] Martin Luther, *Eine kurze Form der zehn Gebote, des Glaubens und des Vaterunsers* (1520), in *WA* 7:205.17.

[12] *Catechism of the Catholic Church* (New York: Doubleday, 1994), 19 (31 f.), for this and the following quotation.

which alone is without origin or end. Thus, in different ways, man can come to know that there exists a reality which is the first cause and final end of all things, a reality 'that everyone calls 'God''" (St. Thomas *Summa Theologica* 1.2.3).[13] Besides Thomas who is quoted here, the subsequent text also refers to Vatican I. While this *Catechism* was published after Vatican II, we notice a continuity between Vatican I and Vatican II with regard to the possibility of knowledge of God from nature. What reason is able to know in some ways through nature is attested to by faith. But in the *Catechism* we only read that God can be known through creation, not that he must be known. Therefore the certainty of Vatican I and II is relativized to a possibility.

It is an open secret that the *Catechism* in its final form bears the mark of Pope emeritus Benedict XVI, when he was still Prefect of the Congregation for the Doctrine of the Faith. In his inaugural sermon of April 24, 2005, Pope Benedict stated: "We are not some casual and meaningless product of evolution. Each of us is the result of a thought of God. Each of us is willed, each of us is loved, each of us is necessary."[14] While this is a conclusion based on theological presuppositions, a statement of April 6, 2006, goes much further. In preparation for the 21st World Youth Day, Pope Benedict said in a meeting with young people from the Diocese of Rome: "Our knowledge, which is at last making it possible to work with the energies of nature, supposes the reliable and intelligent structure of matter. ... The more we can delve into the world with our intelligence, the more clearly the plan of Creation appears."[15] This is clearly a statement which goes beyond a strictly theological assertion. That an intelligent design of some sort is very much on his mind can also be seen from the Pope's meeting with his former students. After he was elected Archbishop of Munich, he still had a number of doctoral students left from his days as professor at the University of Regensburg. He

---

[13] *Catechism of the Catholic Church* (34), 20.

[14] http://www.vatican.va/holy_father/benedict_xvi/homilies/2005/do cuments/hf_ben-xvi_hom_20050424_inizio-pontificato_en.html.

[15] http://www.vatican.va/holy_father/benedict_xvi/speeches/2006/ april/index_en.htm

met with them on an annual basis, usually at Spindelhof, a retreat house near Regensburg. From 2005 and his election as Pope, Cardinal Ratzinger continued these meetings at Castel Gandolfo. The meeting of 2006 (September 1 – 3) was devoted to "Creation and Evolution." As Cardinal Schönborn has pointed out, he was a participant for at least twenty-five years of the so-called "Ratzinger Circle." If the topic of such a meeting is "creation and evolution", this shows that the results of the natural sciences are seriously considered as it is also the case with the cooperation between the Papal Observatory and *CTNS* in Berkeley. One will perhaps be happy to note that with some scientists and philosophers a change of mind is taking place and that not all of the representatives of the theory of evolution reject categorically the mention of a design in nature.

9.2 Brain and Spirit

An area in which research has made great progress in the past few decades is the brain. Austrian neurologists even named the year 1999 as "The Year of the Brain." Since the brain is the main place of integration for all complex processes in the human body and it is in the brain where one receives even better insight into those processes, it is especially the intellect, consciousness, and will which are theologically important. Are mind and consciousness only functions of the brain, as some researchers in evolution claim? Is it only our imagination that tells us we have a free will or are we actually free? If there is no freedom of the will then we are also not responsible (before God) for our actions. In legal terms a distinction must be made between accountability and non-accountability. Yet statistically seen most people who get in trouble with the law are fully accountable.

9.2.1 Mind, Consciousness, Free Will, and the Brain

Concerning the relationship between mind and brain there are three basic theories, a dualistic, a pluralistic, and a monistic one.

In the dualistic theory mind and brain are regarded as separate entities which can enter into connection with each other but cannot be reduced to one side or the other. In the pluralistic theory there exist separate and distinct realities such as the subjective world of the spirit, the physical world of the brain, and the objective world of scientific knowledge. Finally the monistic theory claims that mind and brain are essentially one and the same and that physiological processes can be explained through neurological events in the brain. In spite of these different theories "the ultimate explanation of both mental life and behavior must be sought" within the complex organ of the brain.[16] This view, referred to as the psycho-neurological identity hypothesis, asserts that mental processes and brain processes are identical. Without the brain there is no mind and mental events are processes of the physically working brain. Yet this view need not lead to monism because the whole need not be just the sum total of its parts. Even if we cannot determine the human spirit as being in some respects independent of the brain this does not mean that it is only part of the brain or its functions.

To understand human behavior neuroscience, like any other science, must proceed reductionistically to divide complex phenomena in ever more simple underlying mechanisms which cause these phenomena. If the processes in the brain are severely disturbed through disease, manipulation, or accidents it is easier to correlate activities of the brain or its malfunction with a certain behavior. If those disturbances do not exist such correlation is much more difficult to establish. Moreover we must remember that the brain consists of at least fifteen billion interconnected nerve cells. Each of these cells has up to several thousand synapses which provide important points of contact with other nerve cells. Therefore an exact description of the brain and a complete evaluation of its activities are virtually impossible and we cannot exactly predict how the brain will act in a certain case.

Yet there has been one successful instance. Already in 1965 the German neurologist HANS HELMUT KORNHUBER (1928 – 2009)

---

[16] So Jackson Beatty, *Principles of Neuroscience* (Madison: Brown & Benchmark, 1995), 4.

published results of experiments in which he investigated the electric potential of the brain before and after one's hand was moved. He discovered that for nearly an entire second there is a build-up of brain activity, the so-called readiness potential, before the person who is examined conducts the conscious decision to move a finger.[17] The American physiologist BENJAMIN LIBET (1916 – 2007) refined in 1979 these experiments to detect 1. When does a person decide to move his hand, 2. When does the brain prepare itself for that activity, and 3. When is this activity executed. He noticed that the brain starts to work before a person (consciously) decides to move the hand. This would mean that the brain acts independent of our will. He also conducted experiments in which he touched a person with a needle and noticed that the effect of the touch reaches the brain very quickly but only after a half-second of brain activity the touch is consciously realized.[18] This delay would indicate the possibility of a conscious reaction to the touch, because "a quick reaction to a stimulus … would have to be made unconsciously."[19] Yet a conscious reaction would point toward a free will. Yet one should be very careful with such assertions because the experiments by Kornhuber and Libet were reductionistic since they took place in the laboratory and not in a natural environment which is much more complicated than laboratory conditions. Therefore the issue of a free will is still not clearly decided.

In order that an activity is voluntary it must be presupposed that one could also decide to perform some other action.[20] The

---

[17] L. Deecke/H. Eisinger/H. H. Kornhuber "Comparison of Bereitschaftspotential, Pre-motion Positivity and Motor Potential Preceding Voluntary Flexion and Extension Movements in Man," in *Progress in Brain Research*, vol. 54, ed. H. H. Kornhuber /L. Deecke (Amsterdam: Elsevier/ North-Holland Biomedical Press, 1980), 171 – 176.

[18] Cf. for the following Benjamin Libet, "Subjective and Neural Time. Factors in Conscious Sensory Experience, Studied in Man, and Their Implications for the Mind-Brain Relationship," in John Eccles, ed., *Mind and Brain. The Many-Faceted Problems*, 2nd ed. (New York: Paragon, 1985), 187.

[19] Libet, "Subjective and Neural Time," 188.

[20] Cf. for the following R. E. Passingham, *The Frontal Lobes and Voluntary Action* (Oxford: Oxford University, 1993), 1 f.

activity should not simply be a reaction. Moreover it should occur with attention that is directed to it and not in an automatic fashion as, for instance, it occurs when we shift from one gear to another in our car. Furthermore we should be able to compare this action with others and not habitually do one or the other. But do we have such freedom? Humans and animals have the freedom to adjust to new circumstances. Yet with animals this adjustment occurs through trial, for instance as to how they can obtain food. Humans however can already beforehand contemplate which possibilities are open to them and then proceed in a strategic manner. This means we can choose among various ideas to plan our future activities by setting goals and developing strategies to reach them, thus proving that we are able to act freely or voluntarily. The free will is moreover a subjective phenomenon of our conscience. The poet Friedrich Schiller (1759 – 1805) for instance wrote: "A human is created free, even if he were born in chains."[21]

In more recent times a consensus has developed that consciousness can also be scientifically investigated.[22] Though a complete understanding of the neurological basis of consciousness is not yet possible it has become clear that consciousness is no separate faculty, "but a certain activity, a mode of basic cognitive functions—memory, perception, action planning."[23] But there is not one proof for a single integrated anatomic structure in the brain which functions as the head of a certain hierarchy of processors. "Human consciousness has special memory retrieval features that sets it apart from animal consciousness; it can voluntarily recall items from its own memory store … without needing further specific cues from the environment."[24] Only in bird vocal mimicry and in the use of symbols by acculturated apes do we find such a limited voluntary recall.

---

[21] Friedrich Schiller, "Die Worte des Glaubens", stanza 2.

[22] So Jean Delacour, "An Introduction to the Biology of Consciousness," *Neuropsychologia* (July 1995), 33:1061.

[23] Jean Delacour, "An Introduction to the Biology of Consciousness," 33:1070.

[24] Merlin Donald, "The Neurobiology of Human Consciousness," *Neuropsychologia* (July 1995), 33:1090.

Neuroscience also shows us that in our conduct we are not so free that we can do whatever we want. First we are determined by our psychic constitution. If we know we can influence this constitution to certain degree then we are able to change ourselves within certain limits. We are determined by that which confronts us outside and inside our body, for instance the drinking of alcohol and then descending a stairway. We can influence these factors through the knowledge of these things (we descend the stairs consciously very slowly knowing that we have consumed alcoholic beverages). Yet what does this mean for our free will? It means that within certain limits we have the freedom to reach a certain goal which we have set. But freedom does not mean that there is no causality at work. Nevertheless we evaluate and determine a possible activity according to our own knowledge and our own will. Rational, emotional, and cognitive elements come together.

Neurobiology does not abolish the freedom of the will by associating brain activities with certain conscious behavior but it does show us that our freedom is not without limits and not without a neurological foundation on the basis of which freedom is possible. This means that there is no reason for a "physicalism", a concept which entails that a human being is only a complicated biological apparatus. As the philosopher Nancey Murphy and the neurophysiologist Warren S. Brown state: through brain research the cognitive neurosciences "help us to understand *how* humans succeed in acting reasonably, freely, and responsibly."[25] As finite beings who are subjected to inner and external limitations perhaps we cannot fulfill the double commandment of love, but we can use the results of the neurosciences to minimize these limitations. Since the neurosciences show the connections between physical-chemical processes and a certain behavior we can use these insights to responsibly influence our own behavior.

---

[25] Nancey Murphy/Warren S. Brown, *Did My Neurons Make Me Do It? Philosophical and Neurobiological Perspectives on Moral Responsibility and Free Will* (Oxford: University Press, 2007), 3. For the discussion of "Physicalism" see *ibid.*, 7 f.

9.2.2 Religious Consciousness

Since humans are conscious of God and their experiences with God, this consciousness can be related to certain activities in parts of the brain. But it is too simple to claim that faith or religion is only an activity of the brain. It has long been noted that patients with certain brain disorders, for instance schizophrenia, often exhibit a striking intensification of religious experiences without external stimulation. This provides neuroscientists with the possibility "to investigate neural correlates of religion and spirituality."[26] Yet so far it can only be shown from data provided by neuroscience, which brain region is likely unconsciously impacting conscious experience. It cannot, however, designate "a particular region as the God spot, nor can we determine which specific brain region produces religious or spiritual tendency."[27] The neurologist PATRICK MCNAMARA (*1956) states similarly: "As far as I can see none of the extant cognitive or neuroscience models of the human mind/brain can adequately account for the range of behavioral and cognitive phenomena associated with religion."[28] This means that religious experience has something to do with the brain but does not come from the brain.

In the last few decades, however, new functional neuro-imaging techniques have brought new insights on how the brain supports religiosity, for instance what brain activity occurs when we pray or meditate. The interesting conclusion is that the segments of the brain which were identified as crucial for religiosity with clinical patients "also appear consistently in neuro-imaging findings of healthy persons performing religious practices. ... There is a network of brain regions that consistently are activated when

---

[26] So the psychiatrist Stephan Carlson, "The Neuroscience of Religious Experience: An Introductory Survey," in Volney P. Gay, ed., *Neuroscience and Religion: Brain, Mind, Self, and Soul* (Lanham, MD: Rowman & Littlefield, 2009), 165.

[27] Carlson, "The Neuroscience of Religious Experience," 168.

[28] Patrick McNamara, *The Neuroscience of Religious Experience* (New York: Cambridge University, 2009), x.

persons perform religious acts."[29] This means there are certain areas in the brain that are activated in sick persons to produce religious experiences that are also activated in healthy persons when they perform religious practices. Does this mean that these practices are induced by these brain regions? This would be the case if certain neurotransmitters that are crucially involved in religious experience, such as serotonin—LSD as a drug acts on the serotonin system as well as does mescaline—and dopamine were not considered. The level of these neurotransmitters is higher in religious exercises and religious experiences, and too high with certain brain disorders. Therefore religious experiences in healthy persons are then reflected on the level of these transmitters and in the activities of certain brain regions.

The neurologist ANDREW B. NEWBERG (*1966) has even advocated a neuro-theology for which he sees four basic goals:

1. To improve our understanding of the human mind and brain;
2. To improve our understanding of religion and theology;
3. To improve the human condition, particularly in the context of health and well being;
4. To improve the human condition, particularly in the context of religion and spirituality.[30]

The dialogue between theology and the neurosciences has to continue to come closer to these goals. Yet the dialogue with those who in a reductionistic way advocate a strictly monistic view and perceive mind and consciousness as purely neuro-chemical processes should not be forgotten. In the conversation with them they must understand that their way is not the only way of interpreting neuro-scientific insights. Since this science is making such fast progress new data must also always be examined to determine its theological relevancy.

---

[29] McNamara, *The Neuroscience of Religious Experience,* 127.

[30] Andrew B. Newberg, *Principles of Neurotheology* (Burlington, VT: Ashgate, 2010), 18.

9.3 Responsible Shaping of the World

When we discussed free will we already touched on the subject of human responsibility. This responsibility becomes especially urgent as soon as scientific insights are put into praxis for instance in the technical realm or in the medical field.

9.3.1 Technology between Ability and Permissibility

In 1966 the American government selected the Boeing Company to build a prototype of a commercial supersonic airplane. Yet just four years later the government stopped its subsidy before even the first prototype was built. While Great Britain and France put with the Concorde the supersonic passenger plane into service the U.S.A. denied its possibility. To my knowledge this was the first time that a technological possibility of progress was not put into practice. Since then the issue has re-surfaced again and again as to whether we should be allowed to realize that which we are capable of doing.

We have already now transformed our little planet earth on which we live in such a drastic way that the Dutch chemist and Nobel laureate PAUL CRUTZEN (*1933) claimed with much approval that since the 18th century a new geological epoch has started the anthropocene. He writes: "Unless there is a global catastrophe — a meteorite impact, a world war or a pandemic — mankind will remain a major environmental force for many millennia. A daunting task lies ahead for scientists and engineers to guide society towards environmentally sustainable management during the era of the Anthropocene. This will require appropriate human behavior at all scales."[31] This means we are confronting unprecedented challenges which we have caused through the technological transformation of our scientific knowledge.

For instance in German agriculture there is an ever stronger process of concentration which is accompanied by the so-called

---

[31] Paul J. Crutzen, "Geology of Mankind," *Nature* (January 3, 2002), 415:23.

"dying of farmers." While in Germany this tendency has not yet had dangerous consequences because farmers can often find employment in other professions it is different for the so-called Third World.[32] There many families live exclusively from the yield of their land. If they and their little farms are displaced by agro-business which usually produces for people in industrial countries such luxury products as pineapples, bananas, flowers or "cheap meat" then they no longer have a means to make a living. Here we are confronting the issue of social justice but also of a sustainable use of the land. While small farmers are dependent on their land, agribusiness can transfer their production if the soil is depleted of nutrients. Even in Germany a similar problem has arisen through the production of factory style cultivated plants that are transformed into bio-fuel which is subsidized by the government and therefore provides a lucrative business for large companies. Consequently rent for farmland has gone up so much that traditional farmers can no longer afford the rent for the land which they farm for traditional agriculture.

With the tsunami in Japan and its consequences another area has entered public consciousness again, the issue of energy, especially nuclear energy which makes possible our high standard of living. Since our earth is finite the supply of energy on earth is also finite. Here nuclear energy would serve as a transitional energy till renewable sources of energy are put in place. Yet nuclear energy cannot compete financially with solar energy if one figures in all the cost including subsidies and also the final disposal of the nuclear waste and the power plant itself.[33] Moreover both with a larger accident such as in Japan or with the final disposal of nuclear refuse there are longtime consequences which are ethically highly problematic. Yet it makes little sense if one country unilaterally refuses to use atomic energy and at the same time imports atomic energy from other countries. With such an action the potential problems are transferred to a different region. In the long run only renewable energy is tenable which comes from biomass, water,

---

[32] Cf. the conclusions which Ian Barbour draws in his book *Ethics in an Age of Technology* (San Francisco: HarperSanFrancisco, 1993), 114 f.

[33] So Barbour, *Ethics in an Age of Technology*, 129.

wind, and the sun. Yet with biomass, as we shall see, there are already potential problems. The so-called Third World is more affected by these problems than richer countries because there the large-scale cultivation of industrially useable biomass will either affect the subsistence farmers or the biomass is obtained through deforestation. Therefore in the long run we must also consider the environmental cost and the social cost.—Obtaining energy from water power is already fully used in many countries so there is little possibility for an increase in this area.

Energy obtained through huge windmills still opens quite a potential. Yet here the transfer of energy is the problem, because in densely settled areas where one needs the most energy huge windmills cause noise and also do not look very attractive. Therefore the advantages must be weighed over against the disadvantages of building windmills in close proximity to urban areas with the alternative of building them farther away but connecting them to the urban areas with new electric power lines. The problems are similar with solar energy. For instance in Germany the sunshine is not very intensive and the sun does not shine as often as in other countries, as for instance in the countries of Northern Africa. Huge solar installations could be built in the desert areas of Northern Africa and the gained energy could be funneled to Germany and other European countries. But reliable political conditions are a presupposition that Europe could import solar energy from these countries. Moreover one should not forget that energy from wind and from the sun are not continuously produced. Therefore the issue of storing energy must be solved for times when there is no wind or when the sun does not shine.

With these renewable energy resources we are confronted in part with technological and also important political and social problems which require an interdisciplinary and multi-lateral dialogue. It is especially important to reach a societal agreement which is not blocked by egotistic individual interests. In contrast to our present excessive individualization we must come together again as a community in order that our community, which is so dependent on technology, can even survive. The dialogue then must be expanded and cannot be limited to scientists (technologists) and theologians. This tendency is already noticeable in the political arena when appropriate committees of advisors and commissions of experts are assembled which also include representatives from different segments of society who participate to inform and guide the decision-makers. When in conclusion we look briefly at medical science then the necessity of dialogue again becomes evident.

9.3.2 Medicine as a Helpmate for Life

Medicine concerns all of us because sooner or later all of us must visit a medical doctor or be admitted into a hospital. In the field of medicine progress seems to be non-stoppable. Medical advances seem to be made nearly every day. At the same time we are confronted with the fact that healthcare costs are going up and up, and not just for the aging population but for everyone. To combat the sky-rocketing medical costs, for instance in the Netherlands and in Great Britain, from a certain age onward no expensive surgeries such as kidney transplants are performed. Then there is the large area in which medical interventions destroy one's own or the life of another, or at least drastically change it. We could name here euthanasia, organ transplants, therapy for sexuality and fertility, cloning, stem cell research, neurosurgery, and animal experimentation.

Of course one could categorically assert that life is a gift of God which has to be protected under all circumstances and at every age. Yet "we do not encounter life in general, but in particular living beings, the baby born prematurely whose lungs are hardly ready to breathe air; the old man, long demented, who contracts pneumonia; the two candidates for a single transplantable heart; a patient racked with cancer pain asking to end life" and so on.[34] Who decides in an individual case? And according to which criteria does someone get a chance to continue to live while somebody else has to die? A few decades ago most of these cases would have solved themselves because the affected patients died. This

---

[34] So Albert R. Jonsen, *Bioethics Beyond the Headlines. Who Lives? Who Dies? Who Decides?* (Oxford: Roman & Littlefield, 2005), 3.

means "the development of new medical technologies has raised a myriad of questions at the intersection of culture, morality, and the production and application of scientific discovery."[35] This touches also the core of humanity itself: What is a human being? When does human life begin and when does it end?

For nearly two decades committees and consultants have worked in U.S. hospitals to explore what is meant for instance by the patient's "living will" to do "everything" to keep that patient alive. Given the full arsenal of possible treatments one must decide what is appropriate and what is beneficial. Even when family and physicians disagree a solution must be found. In 1978 a congressionally mandated group was formed called *The President's Commission for the Study of Ethical Problems in Medicine and Biomedical and Behavioral Research.* The commission worked from 1980 – 83 and made various recommendations with regard to such things health care, human research, life sustaining treatments, defining death, etc. Since the issues have become increasingly complex due to economic factors, information technologies, and medical breakthroughs additional centers have been instituted such as at the University of California at Los Angeles Health Systems Human Resources, a center which is to provide education, service, and research to enhance the practice of medicine for patients, families, professionals and the public.

Ethics committees arose in the early 1970s as physicians, administrators, and lay people wrestled with how to use technology in an ethically responsible way. Roman Catholic hospitals were among the first to form them. Members are all appointed by the hospital and are usually hospital administrators or medical staff. A member of the public and a chaplain may also be included. In response to increasingly complex decisions, many hospitals are adding experts in bioethics and offering more sophisticated training to help members make informed ethical decisions.

---

[35] Barbara Koenig and Patricia Marshall, "Anthropology and Bioethics," in *Encylopedia of Bioethics,* 3rd ed., Stephen G. Post, ed. (New York: Macmillan Reference, 2004), 1:217. The five volumes of this encyclopedia contain a wealth of information on the latest developments and their ethical implications.

Since many of the issues go beyond the expertise of a single individual, also in Europe there these boards and associations have been formed. For instance in 1985 the *European Association of Centers of Medical Ethics* had been formed in which eight German institutions are represented. Yet one can note here hardly any essential participation of theologians. This is different with the *European Group on Ethics and Science and New Technologies* of the European Community in which four of the fifteen members are theologians for the present five-year period (2011 – 2016). In 1986 an *Academy for Ethics in Medicine* was founded with its headquarters in Göttingen. It is an independent forum in which different standpoints and persuasions can be included in the discussion. Among its members are medical doctors, nursing staff, philosophers, theologians, lawyers, and representatives of other professions. It wants to further the public as well as the scientific discussion on ethical issues in medicine, the healing professions, and in health care.[36] Since medicine is important for everyone in a very personal way these three examples show the urgency of a multi-lateral dialogue.

In face of the rapid biotechnological progress the inter-disciplinary discussion is especially important for theologians because in biotechnology the human beings themselves are affected and a decisive question is whether justice is done to a person and whether the dignity of that person is preserved. Here the Christian understanding of a human being as created in the image of God remains important. In applying these various means which modern biotechnology makes available to preserve and enhance life we must ask whether the unique and inviolable human dignity is safeguarded or whether humans are shaped according to our plans and desires.[37] In reaction to the bioethical

---

[36] See the meritorious publication of the Roman Catholic theologian and member of the Academy Karl Hunstorfer, *Ärztliches Ethos: Technikbewälti-gung in der modernen Medizin* (Frankfurt am Main: Peter Lang, 2007).

[37] Cf. Nigel M. de S. Cameron, "Bioethics in Christianity," in *Ency-clopedia of Bioethics*, 1:405, who critically notes "the paucity of Christian resources since the fundamental questions of anthropology that are at stake

and biomedical challenges caused in part by a "medicine of gadgets" and the concomitant shortage of care personnel, besides other various ethical approaches, an "ethics of care" evolved. It originated primarily in feminist writings. For this kind of ethics neither commonly binding moral laws nor utilitarian considerations are decisive, but what most people desire is a personal relationship of trust characterized by love, affection, and compassion.[38] This type of ethics includes many important Christian traits and its advocates find the abstract principles of a distanced and "objective" interchange of doctors with patients as often being irrelevant and insufficient.

When Philip Hefner calls humans created co-creators, created in the image of God to administer God's creation according to God's precepts and to develop it further, he characterizes fittingly our present situation. The decisive question however is what does it actually mean to shape the creation in such a way that it provides an optimum yield for both humanity and creation itself? To bring this issue closer to a solution there not only needs to be an intensive dialogue between theology and the natural sciences, but within the whole of society. Albert Einstein rightly emphasized at a symposium in New York in 1941:

> Science without religion is lame,
> religion without science is blind.[39]

---

in these debates have been comprehensively neglected by theologians and Christian bioethicists alike."

[38] Cf. Tom L. Beauchamp/James F. Childress, *Principles of Biomedical Ethics*, 6th ed. (New York: Oxford University, 2009), 36 f.

[39] Albert Einstein, *Ideas and Opinions* (New York: Crown, 1954), 46.

Index of Names

Abbott, Lyman 75, 78 f., 81
Agassiz, Louis 46, 53, 63, 68 – 72, 75
Alexander, Archibald 31, 49, 61
Altner, Günter 106 – 108, 114, 118
Arnaldez, Roger 14
Averroes 14

Barbour, Ian 8, 108 – 113, 131, 134, 142, 173 f., 223
Barrow, John D. 122, 142 – 145, 210
Barth, Karl 86 – 91, 94 f., 97, 101, 125, 129, 172, 189
Beatty, Jackson 216
Beauchamp, Tom L. 51, 228
Beck, Horst W. 15 f., 18, 118 – 120
Beecher, Henry Ward 75, 77 – 79, 81
Benedict XVI, Pope 214
Benk, Andreas 147
Bennett, Patricia 115
Benz, Ernst 42, 98, 114
Berg, Christian 108
Biedenkopf, Kurt H. 132
Bloom, Paul 156
Bohr, Niels 101, 129, 148
Bonhoeffer, Dietrich 105
Bonnet, Charles 18
Born, Max 193
Boslough, John 138
Bowne, Borden P. 67
Breed, David R. 127 f.

Brooke, John Hedley 115, 117, 129
Brown, Ira V. 78, 216
Brown, Warren S. 216, 219
Brunner, Emil 105
Bruno, Giordano 15, 37, 39
Bryan, Jennings 85
Büchner, Ludwig 23 – 29, 36 f., 74
Bulgakov, Sergii 191
Bultmann, Rudolf 102, 129
Burhoe, Ralph Wendell 126 – 128, 197

Calvin, John 186
Cameron, Nigel M. de S. 227
Carlson, Stephan 220
Carnegie, Andrew 80
Carter, Brandon 76, 142 f.
Cherbury, Herbert of 48 f.
Childress, James F. 228
Clausius, Rudolf Emanuel 29
Clayton, Philip 9, 171, 202, 208 f.
Coleridge, Samuel Taylor 54
Collins, Anthony 49, 100
Collins, C. John 49, 100, 125
Comte, Auguste 66
Copernicus, Nicolaus 13 f., 182
Crutzen, Paul J. 222
Cudworth, Ralph 48
Cuvier, George C. 45 f., 68

Daecke, Sigurd 105
Dana, James Dwight 63, 75
Darrow, Clarence 85

Darwin, Charles   8, 25, 33 – 37, 42,
    44 f., 55 – 60, 62 – 67, 69 – 76, 78,
    80 – 82, 86, 89, 105, 109, 114,
    149 f., 153, 160, 163, 170, 182,
    187, 189, 208
Davies, Paul   123, 133 – 138, 209
Dawkins, Richard   148, 152 – 154,
    156, 160, 166, 185 f., 208
Dawson, J. William   75 f.
de Waal, Frans   192
Deane-Drummond, Celia   123,
    129, 172, 189 – 193
Delacour, Jean   218
Dembski, William A.   148, 164
Derham, William   19 f.
Descartes, René   13, 16
Dinter, Astrid   115
Ditfurth, Hoimar von   114
Donald, Merlin   218
Drees, Willem B.   8, 116, 128, 131,
    144, 170, 172 f., 194 – 197
Dürr, Hans-Peter   147 f., 167 – 170

Ebeling, Gerhard   105
Eccles, John   132, 208, 217
Eibach, Ulrich   199
Eigen, Manfred   115
Einstein, Albert   73, 95, 100, 137 f.,
    147, 193 f., 228
Elert, Werner   15
Ellis, George   117
Engels, Friedrich   23, 28, 42
Evers, Dirk   116, 130
Ewald, Günter   120

Fabricius, Johann Albert   20
Felton, Cornelius C.   66
Feuerbach, Ludwig   31 f., 37 f.
Fiske, John   66 – 68, 70, 74, 79, 81
Flew, Antony   210 – 212
Foster, Frank Hugh   60
Frauenstädt, Julius   26

Freud, Sigmund   153, 158
Fuller, Andrew   53, 181

Galilei, Galileo   13, 15 f., 18, 46
Gifford, Lord   128
Gilbert, Thomas   127
Gilkey, Langdon   7, 10 f., 132 f.
Gitt, Werner   148, 161 f., 166
Gladden, Washington   79
Gogarten, Friedrich   102, 105
Goodenough, Ursula   167, 170 f.
Goodwin, Charles W.   55
Görman, Ulf   116
Grassie, William   121, 124
Gray, Asa   63 – 67, 71, 74 f., 77,
    81 f.
Gregerson, Niels Hendrik   116
Gusche, Edith   114

Haeckel, Ernst   24, 34, 36 – 42, 59,
    67, 74, 106
Hafner, Hermann   114
Hägele, Peter C.   114
Haller, Albrecht von   21, 46
Hamann, Johann Georg   20
Harnack, Adolf von   47, 84
Harrison, Peter   129
Hart, Darryl G.   53
Hartshorne, Charles   110
Hawking, Stephen   133, 138 – 142,
    209
Heap, Brian   117
Hebblethwaite, Brian   63
Hefner, Philip   117, 127 f., 172 f.,
    177 – 180, 202 – 204, 228
Hegel, Georg Wilhelm Friedrich
    31, 34, 36, 202
Heim, Karl   14, 94 – 97, 100 f.,
    104 f., 118 f., 199
Heisenberg, Werner   95, 100, 129,
    139, 147, 167, 174
Herrmann, Wilhelm   47

Hirsch, Emanuel   24, 49
Hodge, Archibald Alexander   62,
   72–76, 90
Hodge, Charles   62, 71–76, 90
Hofstadter, Richard   60, 68, 71
Holbach, Heinrich Dietrich von
   22
Holmstrand, Ingemar   95
Howatch, Susan   129
Howe, Günter   89, 101–103
Hoyle, Fred   134
Hübner, Jürgen   105 f., 115, 118,
   120, 199
Hume, David   50 f., 53, 109
Hunstorfer, Karl   227
Huxley, Thomas H.   56 f., 66, 71,
   74, 78, 187

Idreos, Andreas   129
Ijjas, Anna   172, 193 f.
Irwin, William   56 f.

Jackelén, Antje   9, 116 f.
James, William   52, 66, 75, 80 f.
Jaspers, Karl   7
Johnson, Phillip E.   163
Jonsen, Albert R.   225
Jordan, Pascual   102
Jowett, Benjamin   54
Junker, Reinhard   119 f., 165 f.

Kant, Immanuel   17, 20, 25 f., 39,
   41, 51, 65, 109
Kepler, Johannes   13–16, 18, 46,
   61, 105, 163, 199
Klein, Rebekka   119
Knutzen, Martin   20
Koenig, Barbara   226
König, Franz   20, 114
Kornhuber, Hans Helmut   216 f.
Krolzik, Udo   20
Kutschera, Ulrich   148, 159–161

La Mettrie, Julian Offray de   21 f.,
   52 f.
Lakatos, Imre   181
Lange, Friedrich Albert   26
Laplace, Pierre   21
Le Conte, Joseph   68–70
Leibniz, Gottfried Wilhelm   26, 33,
   37, 51
LeMahieu, Dan L.   51
Libet, Benjamin   217
Liebig, Justus   27 f.
Lindberg, David C.   54, 57
Link, Christian   198 f.
Link, Heinrich Friedrich   33, 199
Livingstone, David N.   53, 73
Locke, John   48–50
Lodge, Oliver   41
Loofs, Friedrich   41
Lorenz, Konrad   114
Losch, Andreas   119
Lubac, Henri de   100
Luhmann, Niklas   114
Luthardt, Christoph Ernst   43 f.
Luther, Martin   15, 167, 179, 213
Lyell, Charles   56

Marshall, Patricia   226
Marx, Karl   42
Mayer, Julius Robert   21, 29, 38
McCalla, Arthur   154
McCosh, James   75–77
McFague, Sallie   130
McGrath, Alister   152, 172, 185–
   188
McNamara, Patrick   220 f.
Miller, Hugh   53
Moleschott, Jacob   23, 28–30, 32,
   36
Moltmann, Jürgen   114, 118, 198,
   200–202
Moore, James R.   54
Morris, Henry M.   163

Müller, A. M. Klaus   102 – 104
Müller, J. B.   40, 103 f.
Murphy, Nancey   123, 134, 181,
    209, 219

Napoleon Bonaparte   21
Nebelsick, Harold   13 f.
Newberg, Andrew B.   221
Newton, Isaac   16 f., 21, 33, 41,
    109, 126
Niebuhr, H. Richard   85 f.
Niebuhr, Reinhold   81, 85 f., 129
Nietzsche, Friedrich   40
Nieuwentyt, Bernhard   18, 20
Noll, Mark A.   53, 61
Numbers, Ronald L.   54, 57, 163

Olmstead, Clifton E.   85

Paley, William   51 – 53, 57 f., 141,
    160, 162 f., 187 f.
Panikkar, Raimon   169
Pannenberg, Wolfhart   102 – 105,
    123, 145, 147, 175, 181, 198, 202 –
    206
Pannill, H. Burnell   66
Passingham, Richard E.   217
Peacocke, Arthur   115, 209
Peters, Ted   124, 131, 157, 173, 179,
    181, 204
Planck, Max   101, 147 f., 167, 193
Polanyi, Michael   94
Polkinghorne, John   117, 123, 130,
    172, 182 – 186, 194, 210
Portmann, Adolf   114
Pusey, Edward B.   54

Rade, Martin   84
Ramm, Bernhard   125 f.
Rauschenbusch, Walter   82 f.
Rendtorff, Trutz   120
Rhine, Joseph B.   114

Ritschl, Albrecht   47, 84, 195
Rockefeller, John D.   80
Runehov, Anne   116
Rushd, Ibn see: Averroes   14
Russell, Robert John   109, 122 –
    124, 130, 134, 138, 172 – 177, 181,
    210, 212

Sänger, Eugen   114
Santmire, Paul   181
Schäffer, Jakob Christian Gottlieb
    19
Scherer, Siegfried   120, 148, 161,
    164 – 167
Schiller, Friedrich   218
Schleiermacher, Friedrich   28, 47
Schmitz-Moormann, Karl   105
Schönborn, Christoph   212, 215
Schopenhauer, Arthur   26
Schrödinger, Erwin   148
Schultze, Victor   44
Schütt, Hans-Werner   16
Schwarz, Hans   9, 63, 118, 154,
    200, 212
Schweitz, Lea F.   127
Scopes, John   60, 85
Shapere, Dudley   17
Smedes, Taede A.   116
Spencer, Herbert   37, 58 f., 63, 66 –
    68, 71, 74, 78, 80 – 82, 86
Spinoza, Baruch   37, 39 – 41
Stoeger, William R.   133 – 135, 209
Strauss, David Friedrich   34 – 36,
    39, 74
Sumner, William Graham   80

Tanne, Markus   19
Teilhard de Chardin, Pierre   42,
    97 – 100, 105 f.
Teller, Edward   167
Temple, Frederick   55, 124
Temple, William   55, 124

Templeton, John Marks, Jr.   117, 122
Templeton, John Marks, Sr.   117, 122
Thomas Aquinas   190, 212 f.
Thorpe, Vanessa   156
Tillich, Paul   105
Timm, Hermann   101
Tipler, Frank   122, 138, 142 – 147, 156, 162, 204, 210
Tödt, Heinz-Eduard   102
Torrance, Thomas F.   88 – 94, 185
Trigg, Roger   130

Ullrich, Henrik   119

van Huyssteen, Wentzel   130, 198
Virchow, Rudolf   41 f.
Visser't Hooft, Willem   102
Vogt, Carl   23, 27 f., 33 f., 36 f., 74
von Balthasar, Hans Urs   191

Wagner, Rudolf   27 f.

Wallace, Russel   74
Watts, Fraser   117, 129
Weber, Otto   102
Weizsäcker, Carl Friedrich von   7, 101, 103, 132, 167
Welker, Michael   198
Whitcomb, John C.   163
White, Andrew D.   60, 81
White, Edward A.   81
White, Lynn   81, 106
Whitehead, Alfred North   94, 110, 136
Wiegrebe, Wolfgang   21
Wilberforce, Samuel   56 f.
Wilson, Edward O.   148, 154 – 159
Wittgenstein, Ludwig   137 f.
Wolf, Ernst   101, 103
Worthing, Mark   205
Wright, George F.   75, 77
Wuketits, Franz M.   148 – 152

Zöckler, Otto   33, 42, 44 – 47, 89

Index of Subjects

AAAS 126

American Scientific Affiliation 125 f.

anthropocene 222

anthropomorphism 38, 204

apologetics, apologetic 49, 76, 89, 185

atheism, atheist, atheistic 18, 22, 26, 48 f., 52, 73 f., 148, 152, 185

biotechnology 190, 227

book of nature, *see also* two books 18, 46, 55, 61, 188, 201 f.

Brights Movement 152

British Humanist Association 152

Center for Religion and Science 117, 127

Christians in Science 130

clockwork 140

co-creator 177 f., 180, 208, 228

consciousness 39, 67, 103, 137, 208, 215, 218, 220 f., 223

contingency, contingent 93, 203 – 205, 213

cosmology 124, 133 f., 138, 143, 145 f., 148, 174 – 177, 193, 207, 211

creationism 85, 119, 126, 133, 149, 159 – 164, 166

CTNS 109, 122 – 124, 133 f., 173, 181, 198, 215

deism, deist 21, 48, 53, 142, 146, 197

design, *see also* intelligent design

Discovery Institute 163, 165

dualism, dualistic 22, 36, 93, 98, 111, 208

emergence 16, 94, 149, 178, 185, 187, 208 f.

empiricism 46, 48 – 51, 157

environment, environmental 7, 33, 75 f., 105, 107, 112, 158, 168, 178, 181, 197, 217 f.

eschatology 104, 174 – 176, 184 f.

ESSSAT 115 – 117, 119, 130, 194

ethical, ethics 30, 49, 65, 68, 109 f., 120, 130, 167, 181, 199, 201, 226 – 228

ethics committees 226

evolutionary process 30, 34 f., 42, 58, 70, 72, 76, 80, 178 f., 188

FEST 118, 120, 199

field (theory) 42, 52, 88, 90, 109, 111, 117 f., 120, 129, 134, 154, 176, 204 f., 222, 225

freedom 15, 26, 41, 51, 54, 86, 92, 137, 178, 180, 194, 205, 210, 215, 218 f.

fundamentalism, fundamentalist 85, 126, 152

gene, genetic 152 – 156, 177, 185

Genesis, genesis   29 f., 32, 45, 54,
    62, 79, 125, 162 f., 167
Gifford Lectures   7, 111 f., 128,
    131, 142, 188, 200
Giordano Bruno Foundation   149,
    159

Ian Ramsey Center   129
idealism   36
immortality   29, 41, 51, 142, 144 f.
innate ideas   49
Institute for Creation Research
    163
intelligent design   119, 150, 160 –
    166, 212, 214
Intelligent Design Movement
    163 – 165, 211 f.
IRAS   127 f., 170
ISSR   117

Karl Heim Society   118 – 120

LSTC   127

materialistic, materialism   23 f.,
    27, 30 f., 33, 42 f., 47, 52, 67
mechanistic   22, 52 f., 141, 168
meme   152 f.
metaphysics   12, 50 f., 111, 193,
    210
monist, monistic, monism   31, 42,
    153

naturalism, *see also* naturalism
    22, 81, 121, 153 f., 165 – 167,
    170 f., 196 f.
naturalist   28, 37, 45, 58 f., 63, 68,
    74 f., 77, 196
naturalistic, naturalism   21, 152,
    160, 195, 197

natural selection   35 f., 56, 58, 64,
    70 – 72, 74 – 76, 80, 149, 152,
    155 f., 187
neurobiology   199, 218 f.
nuclear (energy)   11, 106, 113,
    167 f., 223

omega point   98, 144 f.
origin of life   24, 32, 134, 161, 166

panentheistic   209
pantheism   39, 41, 43
physicalism   208, 219
physico-theology   16, 19 – 21, 51
Plan, *see also* design   18, 25, 73, 76,
    136 f., 153, 214, 218, 222, 227
postmodernity   189
process   21, 29 f., 37, 40, 43, 64, 72,
    75, 79, 83, 86, 93, 99 f., 106, 109 –
    113, 136, 144, 149 f., 155 f., 159,
    161, 166, 174, 176 – 179, 183, 188,
    194, 197, 203, 208 f., 215 f., 219,
    221 f.
progress, progressive   8, 16, 21, 26,
    33, 41 – 43, 50, 62 f., 70, 81, 84,
    86, 103, 107, 112 f., 127, 130, 133,
    140, 150, 206 f., 210, 215, 217,
    221 f., 225, 227
*Protestant Study Community, see
    also FEST*   118
purpose, purposeful   53, 72 f., 75,
    112, 117 f., 127, 158, 171, 178,
    182 f., 211 f.

quantum mechanics   73, 96, 143,
    177, 193 f.

reductionistic, reductionism   136,
    145, 217, 221
Reformation   15, 87, 172
relativism   189

religious experience   81, 111,
    220 f.
Roman Catholic (Church)   13,
    96 f., 100, 105, 121, 130, 147, 161,
    172, 189, 191, 193, 198, 212 f.,
    226 f.
RSNG   121

Scottish Realism   90
Second Vatican Council (Vatican
    II)   98
singularity   138 f., 141, 146
social Darwinism   40, 60, 68, 71,
    80, 82
Social Gospel movement   82
soul   17, 21 f., 27, 39, 43, 48, 51, 72,
    99, 125, 144 f., 151, 208, 220
spontaneity   73, 209
*Study Community Word and
    Knowledge*   119, 161, 165
supernatural   65, 72, 79, 128, 136,
    157, 161

technological progress   112, 168
technology   7, 11, 90, 107, 109 –
    113, 119 – 121, 124, 126, 144, 161,
    180, 198, 200, 222 – 224, 226
teleological, teleology   36, 44, 50,
    73, 187, 211
Templeton Foundation   121 f.,
    124 f., 129

Templeton Prize   109, 113, 122,
    127, 134 f.
theism, theistic   41, 65 – 67, 170,
    194
theology   7 – 13, 15, 17 – 19, 23,
    42 – 47, 51 – 57, 60 – 63, 71, 73,
    76 – 80, 83 – 96, 99 – 102, 105 –
    130, 132 – 134, 145 – 148, 154,
    156 f., 167, 169, 172 – 176, 179 –
    191, 193 – 204, 206 f., 212 f., 221,
    228
thermodynamics   21, 29
thought   20, 24, 26, 30 f., 33 f., 36 –
    38, 41 – 43, 48, 50, 60 f., 63, 66,
    68 f., 71, 73 f., 76 – 78, 81, 90, 92,
    96 f., 100 f., 111, 119, 130, 136,
    138, 149, 153, 159, 170, 184,
    186 f., 194, 199, 214
transhumanism   192
TTN   120, 198
two books   55

Vatican Observatory   123 f., 133
Victoria Institute   129 f.
virtue   48, 70, 156, 190 f.
vitalism   41

watchmaker   52, 141

Zygon   9, 63, 92, 117, 127 f., 154,
    170, 177, 192, 194, 197, 204 f.